1. Deer at Glensaugh Experimental Farm. This is the type of land on which they run.

Farming the Red Deer

The first report of an investigation by the
Rowett Research Institute and the Hill Farming Research Organisation

K. L. Blaxter, R. N. B. Kay and G. A. M. Sharman
Rowett Research Institute

J. M. M. Cunningham and W. J. Hamilton
Hill Farming Research Organisation

With a foreword by The Viscount of Arbuthnott
Chairman, The Red Deer Commission

Department of Agriculture and Fisheries for Scotland

Edinburgh Her Majesty's Stationery Office

First published 1974

Second impression 1975

Cover illustration
from Thomas Bewick: A General History of Quadrupeds

ISBN 0 11 491183 5

Contents

List of Illustrations

Foreword

The Hill Farming Research Organisation experimental farm at Glensaugh is on the fringe of an ancient and royal deer-hunting park where the needs of an itinerant court for a food larder probably always outweighed the mere pleasure of a handy chase! The custom of keeping deer in comparatively close confinement in parks as a ready supply of meat has been followed for centuries in many parts of Europe where some deer managers presently engage in the control of their deer stocks, by using winter feeding with or without stockades or corrals. This comes close to a form of husbandry and deer farming itself is not a new concept for it is practised after a fashion in other parts of the world, notably New Zealand.

What is new and challenging about the Rowett/H.F.R.O. idea is the proposition that there might be, in the native red deer of Scotland, a hitherto unexplored, and certainly underexploited, capability for forage conversion and that if the animal were domesticated and viably utilized, it could open up a new dimension to the productivity of the Highlands.

It might have been expected that I and my fellow members of the Red Deer Commission would be sceptical. We did sound a note of caution on practical grounds but also gave the proposal encouragement as a project in two successive annual reports after it was first mooted, for the reason that more research on red deer was certainly needed and that anyone interested in deer should maintain a flexible attitude towards any new enquiry of this sort.

My personal belief that the research would be for the wider benefit of progressive deer management should be vindicated by this report of the work of the research team during the first years of the project from 1970 to 1974. Not only does it confirm what is apparent to any visitor to the study area, that the deer farming enterprise is a thoroughly workmanlike and most professional experiment in animal husbandry, but also that the approach to their problem by the whole team is as competent as it is dedicated and as resourceful as it is pragmatic. Above all, this interim report is the model of concise explanatory recording of the progress of a research project, that is often sought in vain by those of us who encourage research work!

Dr. Blaxter and his team have written a fascinating account of the early years of this unique experiment. They have underlined the many problems that will still need to be overcome and never minimize the difficulties that surround the creation of a deer herd that is sufficiently domesticated to be handled safely and efficiently. They recognize the high costs that would be entailed if adequate domesticity could only be achieved by hand-rearing captured or farm-born calves in their first winter. Their attempts to solve this problem and the economic and practical consequences of failure are well described. An answer here may well be the key to success but this hazard will always be a threat until a race of deer stockmen has been trained to the requirements of deer farming.

Sheer good stockmanship is a vital ingredient in the success of any livestock enterprise and this quality is demonstrated both in the management of the captured animals at Glensaugh and in the early breeding success. The fact that some initial disappointments were overcome, that other likely hazards were foreseen or quickly circumvented and that a realistic appreciation of the livestock management difficulties is so much in evidence, must give even the most critical reader confidence in the ability of this team to reach sound conclusions. The investigation into the veterinary aspects, the health and disease factors, the work on day-length effects on animal performance and even the quest into the palatability of venison have probable relevance far beyond the scope of the deer farm experiment alone and are fascinating studies in themselves.

Finally, in perfecting their fencing techniques and handling facilities they have already demonstrated a satisfactory method of managing captured deer in farming conditions and can now project a tested, and costed, paddock fencing and pen lay-out for the minimum size of deer farm judged by sound stock-handling principles. The probable, but then unknown, cost of fencing was one of the first criticisms levied at the notion of deer farming. Now we can be on firmer ground in asserting that a minimum fencing expenditure of about £7000 (at 1974 prices) per square mile of deer farm area is likely to be an accurate estimate and that 600 acres or thereby is a sensible minimum acreage for such a venture. This is a practical and very worthwhile ancillary result of the present research work.

There will be those who would have ethical grounds for criticism of any project that entails the domestication of a wild animal. There may indeed seem to be good reason to doubt the desirability of promoting the 'unnatural' use of yet another 'natural' resource for the needs of man but such a proposition overlooks the tenable viewpoint that many natural resources in a hard commercial world will only be given their true value and status once it has been demonstrated that they have an intrinsic commercial usability!

If red deer can be seen to compete successfully with other domestic stock on moorland pastures and, ideally, be utilized as an acceptable and economic adjunct to an extensively run red deer forest, is that not itself a better justification for the continued existence of deer forests and wild red deer? In fact one of the limiting factors in the spread of future deer farms may well be a shortage of foundation stock. These can presently best be provided from the wild herd and the two enterprises may best flourish in close association.

As much as anything I appreciated this work for its elimination of at least one of the folklore management techniques of traditional deer forest owners—the need to avoid inbreeding and thus the prevalent belief that introduced new-blood is a necessity! Conversely I might be critical of the lengthy explanation of a calf-catching technique, already so well known and tried by other workers with wild deer, if I did not accept the fact that the method will be unknown to the wider audience to which I hope this excellent report will go.

Judged on the narrow lines of an experiment in animal husbandry the report is a very sound piece of work. Examined as a record of the techniques being tried for the domestication of our largest and finest Scottish wild animal it is a first-class account of the beginning of a most exciting chapter in the extension of the frontiers of knowledge which could well have a profound effect on our understanding of Highland deer and upland land-use.

ARBUTHNOTT

Red Deer Commission,
82 Fairfield Road,
Inverness, IV3 5LH

July 1974

Introduction

This book gives an account of studies commenced by the Rowett Research Institute and the Hill Farming Research Organisation in 1970 designed to find whether it was possible to devise farming systems based on a wild animal native to Scotland, the red deer. The account is not a complete one for to establish whether any agricultural system is technically feasible, biologically sound and economically viable takes many years. When the objectives are to find ways in which an animal which has never been domesticated can be made the centre of an entirely new type of farming it will be appreciated that much time must elapse before final conclusions can be drawn. Even so, the three and a half years of work reported here suggest that the possibility of farming the red deer is certainly not an outlandish idea, and there seems every possibility that successful husbandry systems can be devised. Whether such expectations are realised will be dealt with in subsequent reports which it is hoped to publish at intervals of three to four years.

The whole project undertaken by the two research institutes is unique in that it represents the first attempt for many thousands of years to domesticate a wild ruminant animal in these islands. It has an additional element of uniqueness about it for it has entailed a wide collaboration between many people with widely differing skills, experiences, and indeed opinions! Not only have many members of the staffs of both the Rowett Research Institute (RRI) and the Hill Farming Research Organisation (HFRO) collaborated with scientists from associated organisations but, in addition, landowners and their employees have helped in procuring calves to stock the farm and by giving advice based on their long experience of red deer in the wild. Much advice has also been received from visitors to the farm set up to examine various problems of deer husbandry. The number of visitors has been restricted so to avoid any possible modification of the behaviour of the deer; even so, during the period under review, over 500 visitors with an interest in and knowledge of deer have inspected the farm and its stock and their comments have been most useful.

The whole project is organised by a small group who call themselves the 'Strathfinella Improvement Society', a name which, redolent of the improvers of the late 18th and early 19th centuries, perhaps reflects the fact that all its members realise that what is being attempted is perhaps as revolutionary in an agricultural sense as were the changes made in Britain by the agricultural improvers of 200 years ago. The group consists of:

Dr. K. L. Blaxter, FRSE, FRS, Director (RRI)
Dr. J. M. M. Cunningham, FRSE, Director (HFRO)
Mr. G. A. M. Sharman, MRCVS, Head, Department of Veterinary Pathology (RRI)
Dr. R. N. B. Kay, FRSE, Head, Physiology Department (RRI)
Mr. J. Eadie, Head, Animal Studies Department (HFRO)
Mr. W. J. Hamilton, Officer-in-Charge at the Farm and member of staff of HFRO.

Mr. W. J. Hamilton has the major responsibility for the experimental farm itself, for the construction of its facilities and for much of the observational study of the deer. His continuous observations are supplemented by information gained at monthly gatherings, when staff from the RRI travel to the farm to help with the work. Such visits are understandably popular in the summer time but less so in winter. The stalwarts of the team are Dr. Kay, Mr. Sharman, Mr. Goodall (RRI), Dr. Pamela MacDonald, (née Arman, Nature Conservancy, seconded to RRI) and Miss Pollock (NERC post-graduate student working at RRI), but other staff, their wives and occasional visitors to the Institute have all been inveigled into helping on occasion.

In the summer and autumn, observations and experimental work have been intensified. University students have been employed by both organisations to assist in this work. Thus entomological study of the head fly, *Hydrotaea irritans*, was carried out in successive summers by Mr. A. Berlyn and Mr. P. Geals of Aberdeen University, studies of herbage selection were made in 1973 by Miss Antonia Ineson and Miss Rachel Mohun of Edinburgh University, while Mr. I. Colville, Mr. D. McGregor and Mr. R. McDougall have worked for periods at the farm either to rear calves or to provide a break for Mr. Hamilton. The catching of calves also involved additional help. The 1971 and 1972 catches on the islands both involved Mr. D. May of RRI with, in one year, Mr. G. Mason, and in the other Mr. and Mrs. Sharman and Mr. Mason. Calf rearing at the RRI was the responsibility of Mr. Mackie Smith.

Specific investigations pertinent to the farm, largely on materials collected there, have been made by both organisations. Thus botanical surveys have been carried out by Mr. I. Colville of HFRO and a detailed study of the status of the heather by Miss S. A. Grant of HFRO. The dissection of carcases and their chemical analysis was undertaken by Dr. V. R. Fowler and Mr. K. Pennie of RRI, the examination of faecal samples for evidence of parasitism by Mr. G. C. Hunter and Miss F. Munro (RRI), and chemical analytical studies have been made by Mr. Goodall and by Dr. J. Davidson and his colleagues (RRI). Studies of serum transferrins were made by Dr. E. I. McDougall (RRI) and the studies of endocrine changes were made by Mr. M. Phillippo (RRI). Photography and preparation of figures for publication have been the responsibility of Mr. I. Robb and Mr. R. Middleton (RRI).

Several outside organisations have helped considerably. A careful watch on the forest area is being undertaken by Professor J. D. Matthews and his staff of the University of Aberdeen, the Animal Health Division of the Department of Agriculture and Fisheries for Scotland, notably Mr. H. Clark, have undertaken all tuberculin tests, while Dr. Pamela MacDonald and Dr. Brian Staines, of the Mountain and Moorland Research Station (Nature Conservancy) undertook most of the observational work relating to the 1973 calving. As referred to in Chapter 10, much of the work designed to assess the disease and immunological status of the deer was undertaken by those highly knowledgeable about the diseases concerned. Thus, the Veterinary Pathology Department, University of Glasgow, determined the antibodies to Leptospira in deer, the Moredun Institute investigated the status of the animals with respect to louping ill and tuberculosis, the Department of Zoology, University of Edinburgh, tested sera for antibodies to *Babesia* and Dr. and Mrs. Williams of Raigmore Hospital, Inverness, tested sera for antibodies to *Toxoplasma*. Brucellosis serology was carried out in the Veterinary Investigation Laboratory, Aberdeen.

The writing of the report was a collaborative effort. Those parts dealing with the description of the farm, the design of its fences and handling facilities, and those relating to the breeding, behaviour, and management of the stock were based on the farm records kept by Mr. Hamilton and were compiled by Dr. Blaxter and Mr. Hamilton. The sections which relate to the various measurements of growth and those studies at the Rowett Institute relevant to the farm project were the responsibility of Dr. Kay, while Mr. Sharman wrote those parts which relate to disease and the perinatal problems of the deer. The Strathfinella Improvement Society have indeed been both authors and editors.

The project has been accepted as part of the British contribution to the International Biological Programme, but the investigations would not have been possible at all without the financial support provided by the Department of Agriculture and Fisheries for Scotland. Their part in the formulation of the whole project is described in Chapter 1 and, in addition, they arranged the publication of this the first report.

Chapter 1. **Genesis and its accompanying vicissitudes**

The background of land and resource use

In Scotland some three quarters of the land area can be classed as rough grazings which consist of mountain, hill, moor and deer forest. The total area is about 12½ million acres which considerably exceeds the 3 million acres of similar land in England and Wales. Afforestation has been increasingly used as an alternative to the farming of sheep and hill cattle on some of this land, for much of it is remote, poor, and not highly productive in a grazing situation.

The hills in Scotland support a population of some 2½ million ewes, this number having been fairly static over the last few years. Some of these hill lands are stocked very thinly; the parish returns analyzed by Cunningham, Smith & Doney (1970), for example, show that 10% of all hill sheep require ten or more acres to support every ewe. Some of these same areas, however, also support wild stocks of red deer. The area of true deer forest is 3½ to 4 million acres only, but the area over which the deer forage is probably about 7 million acres (Red Deer Commission, 1973), that is over one third of Scotland's land area, and the total number of red deer is estimated to be about 195,000.

The hill sheep industry of Scotland has always been a low input one and its output, judged by return per acre, equally low. It has been supported by subsidy payment, notably under the Hill and Upland Sheep (Scotland) Scheme of 1967 and successive Winter Keep (Scotland) Schemes, for many years. The total Exchequer support for hill sheep (excluding those on upland farms) in 1972 was nearly £4 million, having risen from a figure of £3 million in 1968. Every hill ewe in fact received in 1972 a subsidy of about £1·65 (Department of Agriculture and Fisheries for Scotland, 1973). Furthermore, in that same year, hill land improvement grants entailed a further investment in hill land of £2·4 millions under the Farm Capital Grant (Scotland) Act of 1970. The present level of support for the hill sheep industry is part of current United Kingdom policy to maintain hill sheep stocks in the face of what might appear to be insuperable difficulties of maintaining any viable economic enterprise in such difficult physical conditions, and this support, together with an adoption of techniques which have arisen from applied agricultural research designed to augment the productivity of this land, may well prove to be effective in increasing productivity in future years. What is certain is that in the late 1960s there was very considerable concern about the very survival of the industry in its present form.

At that time, surveys of the profitability of hill sheep farming showed that any meagre profit was entirely due to the amount of subsidy support. Indeed, it was estimated, on the basis of careful costing, that 130 to 190% of the net profit of these farms was accounted for by the subsidy payments they received and that interest on tenants' capital was, on average, less than 2%. Economies are difficult to make on hill farms for inputs are low and, understandably, many farmers were in considerable financial difficulties. A few jettisoned the time-honoured and central feature of hill sheep management, namely the daily herding of the flock, to substitute occasional gatherings for essential tasks of marking and castrating lambs and shearing the ewes. The hill shepherd was retired and his place taken by casual labour in an effort to economize. This step virtually meant that on those farms the agricultural skills invested in sheep had reached their lowest ebb. The hill sheep, under such circumstances, is hardly to be called a domesticated animal.

The possibility of exploiting the red deer

At this same time, certain developments were occurring on sporting estates. The prices of venison from estate larders rose steeply from a few pence per pound to more than 20. This was largely the result of a demand for deer meat on the part of countries in Europe and the development of an export trade. Despite an import tariff of 5% levied by the EEC it appeared possible that this new demand, which had been ably exploited by the Scottish Game Dealers Association, was likely to continue despite competition from New Zealand. The monetary return from a red deer carcase had, at that time, become about double that from a hill ewe.

It was known that, under the most extensive systems of grazing on the hill, the production of carcase meat from sheep was only 2 to 3 lb/acre (2 to 3 kg/ha). Approximate calculations made from the stag and hind kills on some of the more densely populated deer forests suggested that the production of venison was probably in excess of 1 lb/acre (1 kg/ha). In addition, hides, antlers and tusks provided by-products for sale. The value of the carcase had certainly become a substantial item in the economy of the estates though still less than the rentals of the deer forests for stalking. It seemed possible that meat production from this poor land might well be similar under some type of deer husbandry system to what it was under a subsistence hill farming economy. It also seemed possible that a wild animal, which for many hundreds of years had survived under these poor conditions on the hills, might be better adapted to them than the domesticated sheep. This latter supposition was supported by a series of investigations of the nature of the diet of the red deer and the hill sheep made at the Rowett Research Institute (1969). They showed that the deer apparently consumed and fermented in its rumen a greater proportion of heather than did the sheep on the same hill. Furthermore, the experience gained in handling captive deer in physiological and nutritional studies showed how very tame they could become (Maloiy, 1968).

The idea of exploiting the red deer to make better use of hill land arose from these considerations. In 1967 the magnitude of the trade in deer meat from wild stocks was not known with any precision but it was estimated that about 25,000 carcases were reaching dealers and that the export trade was close to 1,500 tons/annum. This amount of carcase meat is very small indeed relative to the established meat industry of Scotland. Home production of mutton and lamb, for example, in 1966–67 was 60 thousand tons and of beef 140 thousand tons. The contribution of deer meat was less than 1% of total supplies (Department of Agriculture and Fisheries for Scotland, 1969). The survival of even this small meat industry so dependent on a continental demand was, however, problematic. There were rumours of a likely requirement by German authorities for rigorous inspection of carcases because salmonellae had been found in carcases coming from Eastern European countries. Clearly any meat industry of any scale based on red deer would have little chance of survival if the hygienic standards adopted in handling the carcases were not impeccable, and similar to those applying to domesticated farm livestock. It did not seem likely that an industry could be based on the stalking and shooting of a managed stock; animals killed long distances away from a road, disembowelled where they fell, transported by pony to a collecting point and only later transferred to a meat processing plant could hardly be said to be handled in an hygienic way. For this reason it was decided that if any attempt was to be made to exploit the red deer as a meat animal this exploitation should be based on a partial domestication so that suitable animals could be brought to a central point for slaughter and subsequent processing. An exploitation based on hunting or stalking seemed out of the question.

Initial attempts to finance the project

Once this decision had been taken, further thought was given to the ways in which this might be done. Central and selective slaughter implies a stock that can be handled and herded to be brought to a central point. The stock must therefore be fenced in a reasonable-sized area so that their density is high; they must be individually identified and they must be managed to be made tractable. The cost of the additional inputs is so high that annual production of carcases needs to be considerably greater than in the wild if such costs are to be recouped. However, once such a step of confinement and concomitant management is taken, then other agricultural practices can be applied to increase productivity—notably restriction of the number of intact males, control of diseases and parasites by medication, the provision of additional feed at times of scarcity and the managerial control of the grazing itself. In other words, once the decision to slaughter at a central point is taken, exploitation inevitably leads to the idea of a farming enterprise rather than one of wild-life management.

It was with this basic concept in mind that the Rowett Institute approached the Agricultural Research Council in May, 1967 to request funds for research on deer farming. The Council, however, was not in favour of the scheme, stating that the Institute's main concern should be with animals of known agricultural importance. They asked, however, the very question which had been predicated, namely what contribution could the deer make to the British economy in 15 years' time? Obviously this could not be answered but only guessed. The guess made was that a wild and domesticated stock of 1 million deer was not beyond possibility, with an annual production of 10 to 30 thousand tons of meat per year. An approach was made at the same time to the Department of Agriculture and Fisheries for Scotland. Their reply, while non-committal, was more encouraging than that of the ARC. They pointed out that their general attitude was to some extent reflected in the control provisions of the Deer (Scotland) Act 1959 and the establishment of the Red Deer Commission, namely that deer were incompatible with hill farming unless controlled. Nevertheless, the Department promised to discuss the matter with the Red Deer Commission and also with The Highlands and Islands Development Board. Other approaches were also explored. The Nature Conservancy had consulted the Rowett Institute about the possibility of establishing a deer research unit to develop work similar to that of the Institute on those aspects of the physiology and nutrition of the deer which seemed relevant to the Conservancy's interest. It seemed possible that a joint approach to the commercial exploitation of the deer might follow and this was explored. The Conservancy, however, were primarily interested in the extension of their unique and rewarding studies of the wild animal, its ecology, population dynamics and nutrition rather than in commercial exploitation. It was realized that although the objectives of the Institute and the Conservancy were not the same they had in common an interest in the physiology of the animal and accordingly in 1970 the Institute was pleased to provide working space for Dr. Pamela MacDonald (née Arman) of the Conservancy to study the deer in confinement, while the Conservancy was happy to help the Institute—and later the farming enterprise—by providing both advice and animals.

Continuing its efforts to obtain support and particularly land on which a deer farming enterprise could be started, approaches were made to landowners and the proprietors of deer forests. Interest was aroused and many helpful comments were made, particularly regarding the whole philosophy of marginal land use and the future of the Highlands—a problem which has vexed Scotland since the Clearances and the advent of the 'great sheep'—but no enthusiastic offer of support was forthcoming. Nor was an approach to the office of the International Biological Programme any more fruitful in terms of financial help. However, its Director, Dr. E. B. Worthington, wrote: 'If deer could be herded and handled similarly to mountain sheep much more efficient use could be made of the habitat'; this accorded with the Institute's belief.

In the summer of 1968, more than a year after the first formulation of the project, the early offer by the Department to gain the interest of the Highlands and Islands Development Board became a reality. After a number of discussions at the Rowett Institute, Mr. Robin Dean, projects co-ordinator for the Board, was able, in September 1968, to approach his Board with concrete proposals for a collaborative venture to commence in 1969 based on the Rowett Institute's plans. The initial view of the Board, however, was that being concerned with development they could hardly finance anything which was not likely to be successful and certainly could not provide funds for a research undertaking. The Institute was equally unwilling to state that deer farming was a practical proposition until research had shown this to be so. The two organizations thought that it would be desirable to call a meeting of all interested parties to obtain a collective opinion on the probability of success of what was called the Board's project. This was the origin of the Rowett Deer Conference of January 1969 (Bannerman & Blaxter, 1969).

The Rowett Deer Conference

The Conference—to quote the Red Deer Commission (1970)—'was attended by probably the widest and most knowledgeable representation of all those interested in red deer'. The audience included deer forest owners, owners and wardens of deer parks, foresters, stalkers, hill farmers, scientists and veterinarians, and those connected with organizations which had a direct or indirect interest in the red deer—the Red Deer Commission, the British Deer Society, the Nature Conservancy, the Natural Environment Research Council, the Scottish Landowners Federation, the Agricultural Research Council, the Hill Farming Research Organisation, the Reindeer Council, the Highlands and Islands Development Board and the Department of Agriculture and Fisheries for Scotland.

The proposals for the husbanding of the deer were first put forward by Mr. Dean. The Chairman of the Red Deer Commission then gave their opinion of which the following is a quotation from their concluding views: 'If a fraction of the money which would be involved in the provision of deer fencing for farming were to be used instead to improve the keep on deer wintering ground, this would lead to improved performance by the wild herd and could lead to a more . . . significant improvement in the contribution made by deer to Highland land use. This, with a real stimulus to venison marketing, would seem to be a sounder approach than deer farming in present circumstances'. The Game Exporters Association, however, were in support, Mr. G. H. M. Bland concluding: 'The benefits accruing to marketeers from the farming of deer would be considerable and they would be better placed in the long term further to promote a valuable Scottish food resource'. Other concepts of farming—of game ranching—were put forward to suggest that the scale of the Board's proposal was too small, and the relative importance of stalking rentals and venison sales were duly debated. Even the possibility of a subsidy scheme for deer was raised! Calculations of possible returns were made to show that revenue from a venison enterprise alone would be less than from a hill sheep stock on the same area and that both would be less than the revenue from hind and stag stalking with some recovery of meat, to conclude that any investment was better made with sheep, neglecting perhaps the failure of subsidy support in the past to lead to a thriving industry. The Conference was summed up by Professor Wynne-Edwards, then Secretary of the Natural Environment Research Council and formerly a member of the Red Deer Commission. He made a plea for low-cost range improvement and stated that it was for the Red Deer Commission to provide Her Majesty's Secretary of State for Scotland with advice on management and conservation of the deer and for the Nature Conservancy to provide the scientific advice about the animal which the Commission needed. He rejected the whole idea put forward for husbanding the deer stating that: 'Intensive deer farming would, in practice, be fraught with difficulty and almost certainly unprofitable financially'.

Understandably perhaps the Conference's conclusions were along similar lines. The pertinent resolution was: 'We have considered the Board's specific proposal to increase the production of venison by husbanding deer. We have concluded that such an undertaking, entailing "farming" deer specifically for venison production, if carried out under present circumstances is unlikely to be economic'. This negative view was, however, softened by adding: 'We are strongly of the opinion that investigations of the type envisaged by the Board . . . should be undertaken'—in other words, research should precede development. Other recommendations not relevant to the main issue related to the recreational and sporting aspects of deer management.

Understandably, in the light of this unequivocal recommendation, the Highlands Board reluctantly withdrew its offer of financial support. Those who had had great hopes of at least an opportunity to explore whether partial domestication of deer was possible, and whether sheep—relative new-comers to the hills—were necessarily the best animals to make fullest use of them were very disappointed. Mr. McDonald of the Board, in opening the Conference, had remarked: 'Every new development is a process of experimentation which, like marriage, should be entered into with caution and circumspection'. It appeared that such was the caution and circumspection that the whole concept might never attain a consummation!

After the Conference, representatives of the Board stated that, as the recommendation was based on purely economic arguments, it did not preclude reconsideration of the proposals at some later date. The Board encouraged the Institute to regard the matter simply as 'shelved' to await a more propitious occasion. This was not sufficient for the Institute. The possibility of an even more intensive examination of the potential of the red deer on low land was considered and another approach was made to the Department of Agriculture and Fisheries for Scotland to find that their view was that the Rowett Conference raised a series of questions which required answers and that the Institute, having posed them, was well placed to answer them. The Institute was asked to make concrete proposals, with a view to possible implementation in April of 1970, to be considered by the Department and discussed with the Agricultural Research Council. It was further suggested that a joint proposal by the Hill Farming Research Organisation and the Institute would be viewed with favour. During the autumn and winter of 1969 to 1970 the two Scottish research institutes, the Rowett Research Institute (RRI) and the Hill Farming Research Organisation (HFRO) arrived at an acceptable joint plan. This was approved by the Governing Body of HFRO and submitted to the Department. In March of 1970 the Department gave its approval to the plan and provided funds to start an experimental investigation of the application of agricultural practices to the red deer.

It had taken three years to obtain support for the project and looking back it is apparent that the opposition initially voiced by many was modified and softened by the passage of time and by an increased familiarity with the concepts involved which were so different from traditional ones about wild animals and the Highland environment. The press helped to encourage this familiarity for they gave much publicity to the Conference and its aftermath, so much so that the Institute received many offers of facilities for research during the early part of 1970 when it was already fairly certain that support had been secured. The delay had merit in that it ensured that the project started in an atmosphere of what may be called kindly scepticism rather than vociferous antagonism on the part of many with an interest in the red deer and land economy. It is true to say that as the work has proceeded, kindly scepticism has given way to enthusiastic reference to the potential importance of the researches being undertaken by the two collaborating agricultural research institutes financed by the Scottish Department.

References

BANNERMAN, M. M. and BLAXTER, K. L. (1969) (editors). *The Husbanding of Red Deer: Proceedings of a Conference held at the Rowett Institute, Aberdeen.* Aberdeen: Highlands & Islands Development Board and Rowett Research Institute.

CUNNINGHAM, J. M. M., SMITH, A. D. M. and DONEY, J. M. (1970). Trends in livestock populations in hill areas in Scotland. *Hill Farming Res. Organisation, 5th Rept.*, 88–95.

DEPARTMENT OF AGRICULTURE AND FISHERIES FOR SCOTLAND (1969). *Scottish Agricultural Economics*, **19**, 255–265. Edinburgh: H.M.S.O.

DEPARTMENT OF AGRICULTURE AND FISHERIES FOR SCOTLAND (1973). *Agriculture in Scotland*, 1972. Cmnd. 5255, Edinburgh: H.M.S.O.

MALOIY, G. M. O. (1968). The physiology of digestion and metabolism in the red deer (*Cervus elaphus* L.). *Ph.D. thesis: University of Aberdeen.*

RED DEER COMMISSION (1970). Annual Report for 1969. Edinburgh: H.M.S.O.

RED DEER COMMISSION (1973). Annual Report for 1972. Edinburgh: H.M.S.O.

ROWETT RESEARCH INSTITUTE (1969). Annual Report of Studies in Animal Nutrition and Allied Sciences, **25**, 37–38.

Chapter 2. The planning of the enterprise

The proposals made to the Department of Agriculture and Fisheries for Scotland for a research project on husbanding the red deer were more detailed than those which had been the basis of earlier attempts to obtain funds for they were made in the context of a likely site and with knowledge of the particular problems of climate, vegetation, topography and access that the site imposed. Even so, information on which to make plans was still meagre. While results of extensive studies of the ecology of the wild deer, descriptions of the managerial practices used in ornamental deer parks and records of yields of carcases from shot animals on sporting estates were available, and while limited experience had been gained at the Rowett Institute of keeping five hinds in close confinement, there was virtually no knowledge about the many aspects of the behaviour, breeding, nutrition and health of the deer necessary to plan a major investigation. Planning in the early stages was thus based almost entirely on what was thought, intuitively, to be feasible; indeed, it soon became apparent that an essential feature of the plan would be the incorporation of a resilience so that initial mistakes in planning could be rectified without jeopardizing the overall programme. The project was obviously to be an exploration of possibilities, some of which were not even apparent at the beginning, rather than a test of a particular scheme.

Even with such a loose definition of the detail of the investigations their ultimate purpose had to be rigorously defined. This was to find whether an agricultural system or systems for husbanding the red deer for meat production could be devised which would make as good or better use of marginal or hill land as do the existing alternatives of sheep farming and forestry. In keeping with the initial decisions about resilience in planning no direct comparisons of a deer enterprise with sheep or forestry enterprises were incorporated. It was thought that sufficient information would accrue from the investigations to enable such comparisons to be calculated or, later, to provide the basis for devising a larger investigation of economic and technical feasibility.

Planning thus involved considering what information would be needed and then deciding how best to organize the first few years' work to provide it, realizing that an initial assessment of the relative importance of different aspects might prove to be mistaken or in error. It may be said that in 1973 it is already apparent that some problems thought initially to be of primary importance have not proved to be so and that some problems were encountered that had not been thought about at all in the planning stage. The factors taken into account in planning are discussed under three headings.

Numbers of deer in relation to the resources of the farm

The first group of factors considered in planning the 530 acre farm all related to what may be called the population dynamics of a meat production system based on deer, and estimation of the overall numbers of deer to be managed on the farm. Here general agricultural principles could be used to arrive at estimates. The output of meat from a self-contained stock of animals solely dependent on the herbage resources of an area depends firstly on the number of breeding females and their annual breeding rate, secondly on the number of males that have to be maintained, thirdly on the number of animals which have to be reared to replace female and male breeding stock as these die or are culled, fourthly on the number of animals produced each year which are surplus to these needs, fifthly on the time that these 'surplus' animals take to reach marketable weight, and lastly the final carcase yields of meat. The animal population at any time always consists of breeding females, breeding males, replacement stock being reared and growing animals destined to be killed for meat. This number has to be matched to the feed resources available, and these are at their minimum during the winter months. When the land provides all feed for the animals the main determinant of numbers is its capacity to carry a stock in winter. Supplementary feeding during winter increases this capacity and hence total numbers. Upper limits to numbers are set by disease incidence; the greater the stocking density the greater the risk of disease spread.

These general principles were related to the sparse information available about deer and the carrying capacity of the farm. Mitchell (1969), on the basis of studies of wild stocks of red deer on the island of Rhum, had obtained some data on reproductive rate to show that 40% of mature hinds failed to breed each year, and that the majority did not calve for the first time until 3 years of age. In addition, he found that of the calves born over a quarter were lost through death and these obviously would not contribute to the ultimate yield of replacement or meat animals. Reproduction in deer parks was more successful but when mortality was taken into account only marginally so. It was thought that a farming enterprise might do better; tentatively a calving rate of 90% was used as a basis for planning. Estimates made by stalkers and deer forest owners of the ratio of hinds to stags which would result in successful mating varied widely from 5:1 to 30:1. A narrow ratio was adopted for planning purposes. Estimates of growth rate in the wild suggested that meat animals would have to be kept through two successive winters to provide reasonable carcases. It was, however, hoped that growth rate might be accelerated and that a two-year carry-through of stock might not be necessary. The relative carrying capacity of the farm was assessed from the fact that it had once been stocked with ewes with a land allowance of about the equivalent of 5 acres to the ewe. In the traditional system of sheep management on the farm the ewes were removed from the hill grazings during late pregnancy and early lactation (lambing) and again for 'flushing' and mating in the autumn. The sheep, therefore, grazed the hill for only 8 months of the year, the other 4 months being spent on in-bye pasture. It was thought on this basis that it would be reasonable to build up to a total stocking of about 5 acres to the deer. This, at the time, seemed likely to prove a very heavy stocking of the farm, but it was thought that such a heavy stocking might well provide an earlier solution to some of the problems envisaged. The plans included the provision of winter feed where it appeared necessary and an accounting of this additional input. The final winter stocking of the farm was thus envisaged to be:

Breeding hinds	50
Growing and breeding stags	15
Replacement hinds	10
Calves and meat stock, some of which might be castrated	40
Total deer	115 on 530 acres

Control and management of the herd

The second group of factors considered in planning all related to the devising of ways in which the stock could be managed and handled. The primary one was how to ensure that the stock were sufficiently tame to be herded. Herding is necessary for several reasons. Central to the whole concept of farming the deer is that of central slaughter under hygienic conditions; this means gathering and possibly transport of animals. In addition, an agricultural approach involves the identification of individuals, separation of age classes, movement of animals to different fenced areas so to manage the grazing and to ensure that classes with the greatest need receive the best pasture, collection of them for any routine veterinary treatment, and some measure of control of access to the hinds by the stags at the time of the rut so that constructive breeding policies can be adopted. All these processes entail a controlled movement of animals and some involve individual restraint.

Deer are wary animals and when disturbed have behavioural patterns which vary with the sexes; stags tend to disperse, hinds to move as a group. They do not exhibit a flocking behaviour as do sheep and to a lesser extent cattle. Furthermore, the sexes have distinct seasonal patterns of separation, and stags, particularly at the time of the rut, are known to be savage and dangerous to man. Many charming anecdotes, however, show that deer raised as pets in close proximity to man become very tame and experience at the Rowett Institute with wild adult hinds suggested that their innate wildness and fear of man could be reduced by handling. It seemed possible that by an early handling of the deer calves and by continued reinforcement of the experience which such calves gained by hand-feeding and moving them, a reasonably tame stock could be achieved. It was clearly not desirable to make the stock so tame, so dependent on man, that they had no innate wariness at all; nothing more dangerous can be imagined than a stag in the October-November rut which is so tame that it has no residual fear of man at all.

The decision was, therefore, taken to establish the herd with calves which had been hand-reared. It was thought that this would at least provide an initial tameness and even if tameness persisted over the years only in terms of a familiarity with the home range on which they had been weaned, this would be advantageous. It was realized that the critical question of whether animals reared naturally on the hill could be managed and herded would not be raised until some years had elapsed. Similarly, the building up of a herd balanced in terms of age classes would take some time if establishment was based entirely on the rearing of calves. Even if establishment was spread over two or three years this problem would remain.

Planning also involved the division of the farm into areas and the disposition of these bearing in mind the need to keep the home range contiguous to the handling facilities. These facilities too had to be planned in such a way that animals could be subjected to restraint without fear of injury to them or to those handling them. The facilities at Warnham Park, where there was experience of handling a park stock of red deer, were visited and Captain Lucas gave freely of his advice. The Duke of Bedford's facilities at Woburn were also considered; these had been built nearly 100 years previously when park deer were fattened for meat. In addition, information was obtained from the wardens of United States Game Reserves where much experience had been gained on fence disposition to control movements of elk and where facilities had been erected for handling wild stocks. In addition, the design of sheep handling facilities developed over the years within the Hill Farming Research Organisation were considered. These deliberations resulted in a compromise based mainly on a circular sheep dipping crush which was enlarged to accommodate deer in such a way that the animals could be moved without danger to the operator. The critical dimensions of the race were based on measurements of a large hind penned at the time at the Rowett Research Institute. The lines of fencing and the handling facility are described in detail in Chapter 4.

Facilities for monitoring and investigation

A programme of work designed to allow accretion of information, the nature of which could only be conjectural, rather than to test some simple hypothesis entailed making a fairly wide provision of facilities for monitoring and ensuring that such provision could be changed as occasion demanded. The initial plans for scientific observation were kept as simple as possible but it was kept in mind that different scientific skills would have to be brought to bear as new problems arose. The first plans, therefore, called for a monthly gathering of all stock at which time, body weight and body measurements would be recorded together with any other simple observations which could only be made on confined animals—teeth eruption, hair growth pattern, etc. At these same times clinical examination for external parasites would be made, faecal samples taken for laboratory diagnosis of internal parasitism and where necessary blood would be withdrawn to assess the presence of antibodies to diseases of economic importance known to occur in sheep and cattle. In addition, and relevant to the overall problem of land use, plans were made to assess changes in the botanical composition of the various parts of the farm where the deer would graze or be wintered. This included an assessment of the amount of damage caused by the deer to trees in a small area of established conifer woodland contiguous to the farm and kindly rented to us by the Forestry Commission. All these activities called for visits by scientists to the farm at intervals, some on a monthly basis, others at much longer intervals, rather than for a scientific staff resident at the farm. The two organisations agreed to provide the skilled help in their respective fields of expertise on this basis and at the same time HFRO agreed to appoint a skilled animal husbandman to take day-to-day control of the project at the operational level.

It was envisaged that as work progressed, experiments to reinforce observations or to answer specific questions would be undertaken. In addition, however, the continuing programme of work at the Rowett Institute concerned with the physiological responses of the deer was so arranged that certain questions basic to the operation of the farm could be answered to guide decision-making at the farm. One of the earliest of these experimental investigations under controlled conditions was to determine the voluntary intake of feed and growth rate of calves given feed of high quality. This experiment and others which relate to the progressive development of the farming operation are described in Chapter 11.

Tentative plans were also made to deal with the meat output from the farm. These involved the planning of killing facilities and associated cold storage to be erected near the farm and making arrangements for carcase dissection and appraisal. The marketing of meat, possible outlets for hides and other products of a meat industry based on deer, and questions of what might constitute quality in the venison produced were matters not considered in any detail at this stage, since the first essential was to find whether a meat production enterprise was feasible at all.

The final proposal

The conclusions drawn from the considerations given above were sufficient to form the basis of the proposals made to the Department of Agriculture and Fisheries for Scotland. They were costed using the best available estimates of the expenses to be incurred in setting up the farm and running it. The calculations showed that capital costs for fencing and equipment would be £7200 and running costs would be £5300 for the first two years when the stock was being reared, falling to £3550 per annum later. It appeared that over a 10-year period the enterprise would involve expenditure, additional to the two institutes' normal budget, of £46,000. These costs, estimated in 1969, proved to be reasonably correct, taking into account inflationary changes. Approval of these plans and provisions of funds to enable a start to be made was received on 20th April 1970.

Reference

MITCHELL, B. (1969). The potential output of meat as estimated from natural and park populations of red deer. In *The Husbanding of Red Deer*, pp. 16–22. Ed. by Bannerman, M. M. & Blaxter, K. L. (See references to Chapter 1).

Chapter 3. **The farm**

The area selected for the deer farming project was one of the remoter hill grazings of the Glensaugh farm occupied by the Hill Farming Research Organisation. Glensaugh farm lies across the Highland Boundary Fault on the east foothills of the Grampians in Strathfinella, Kincardine, 12 miles from the sea. It was previously owned by the North of Scotland College of Agriculture and so records of the stocking of the hills and of their general management were available for a period of 30 years. In addition, a number of detailed surveys of the soils of Glensaugh and its vegetation had been made and records of meteorological observations had been collected at the main Glensaugh steading since 1949. Part of the Forest of Drumtochty abutted the area selected and 38 acres of this was rented. Of this 15 acres had been planted in 1937 and was line-thinned by the Forestry Commissioners before being included within the boundary fence of the deer farm. The Forestry Commission also allowed the two organizations access to the farm by a tortuous road through their forests; this indeed provides the only vehicular approach to the farm, and has ensured the essential element of seclusion and isolation that the enterprise required. The description of the farm which follows is largely based on information accumulated at Glensaugh before the farm commenced, supplemented by additional botanical surveys of the hill grazing and forest areas, and other observations made during the period 1970–73.

Topography and climate

Figure 3:1 (derived from the O.S. sheet, Kincardine XVIII SE) shows the general topography of the deer farm. The area is characterized by the deep ravine of the Slochd through which runs

Figure 3:1 The topography of the farm

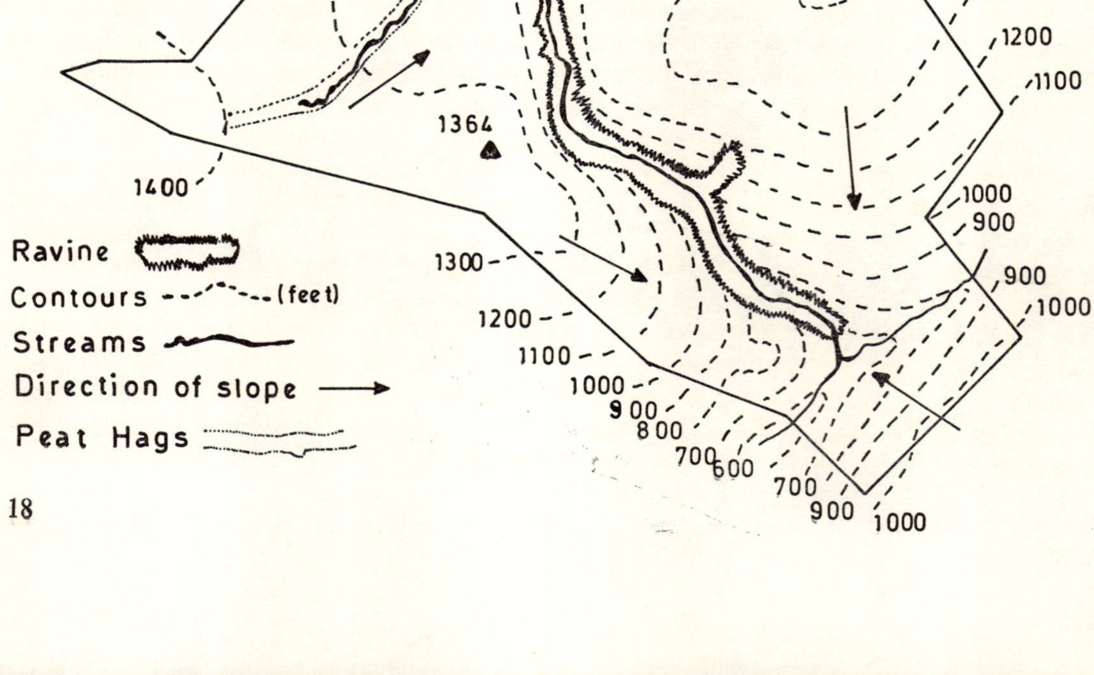

the Slack Burn which, with its feeders, drains the plateau and peat hags of the Waird of the Cairn. A tributary of the Slack Burn, the Burn of Greenshiels, rises in the forest and separates the hill of Greenshiels from the major area. The lowest point of the farm is the confluence of the burns at 600 ft. and the farm rises to its highest point at 1420 ft. to the north-east. Beyond the exposed plateau to the north and north-east lies undulating country on the eroded granites leading to Glen Dye. The farm does not extend to the summit of Greenshiels in the south which is about 1000 ft. and thus that part of the farm in the valley of the Slack and Greenshiels burns is well sheltered from the south. There is evidence of old human habitation in the forest area and downstream from the farm, but little of the residual effects of crofting on the farm itself.

Climatic records made at the steading at Glensaugh are summarized in Table 3:1 for the six years from 1967 to 1972 inclusive. Generally the weather at the steading at 560 ft. shows the area to have an annual rainfall of 35 to 40 inches, March being the driest month and January the wettest. Mean monthly temperatures at the steading fall to 30 to 32°F in winter, February being the coldest month, and rise to 55 to 60°F in July and August. Snowfall in winter can be heavy and the Fettercairn-Banchory turnpike immediately to the east is one of the earliest of Scottish roads to be rendered impassable by snowfall in winter. The farm with its greater elevation and exposure experiences lower air temperatures than the steading, calculated to be 1°C less, while on the exposed plateau to the north air velocity is obviously greatly increased above that estimated in the glen.

Table 3:1 **Rainfall, sunshine hours and air temperature at Glensaugh steading 1967-72**

	Jan	Feb	Mar	Apr	May	June	July	Aug	Sept	Oct	Nov	Dec	Annual values	
Rainfall (inches)	5·6	3·0	2·07	2·5	4·8	2·41	3·02	2·83	2·31	2·8	3·9	2·61	37·7	(Total)
Sunshine (hours)	37	71	105	144	123	147	143	130	90	81	61	39	1172	(Total)
Shade temperature (°F)	37	34	38	42	46	52	55	56	52	48	39	38	44·8	(Mean)

Calculated for the monthly totals for the years 1967 to 1972 inclusive—as gathered at Glensaugh.

Comparison with these average climatic data for the farm shows that the first three winters during which the farm has been operational have been less severe than expectation. Temperatures have been higher and there has been very little snow. Indeed, although a snow tractor had been obtained to assure access to the farm, it has been used for two to three weeks only in the winters of 1970 to 1971 and 1972 to 1973. This atypical weather in the early years of the project may well have proved fortunate; it certainly ensured that none of the stock has required storm feeding for extended periods and that herbage was always available for them. Storm feeding was only necessary in the two short periods already mentioned.

Geology and soils

Lying to the north of the boundary fault the underlying strata are entirely of the Dalradian or Cambrian metamorphic series with igneous intrusions and Figure 3:2 shows their disposition. These rocks are quartz schists and quartz-mica schists (Anderson, 1945; Campbell, 1913). The quartz porphyry intrusion cuts these Cambrian rocks and is particularly well exposed on the scree of the Slochd. The area was at one time extensively glaciated.

The soils of the farm have been extensively surveyed by the Macaulay Institute for Soil Research (1947–48), notably by Glentworth and his colleagues, an account of which was published by Nicholson and Robertson (1958). A map based on their definitive study is given in Figure 3:3, and the original 1943 survey was supplemented by maps prepared by Dr. Glentworth in 1972. The soils are of what Glentworth named the Strichen, Corby and Strathfinella associations. Outstanding from Figure 3:3 is the area of peat to the north which covers the whole of the area above about 1100 ft. This peat, which is of the blanket bog type, is in places up to 4 metres deep but more commonly 2 metres, thinning on the steeper slopes, and is very acid. The feeders of the Slack Burn are associated with small hags but these are not extensive. The peat thins into the podsolized soils which have well developed iron pans, and these are modified by drainage from the peat. More weakly podsolized soils are found at lower levels. These soils are all poor in calcium and phosphorus. In badly drained areas at lower elevations gleys are found, on the till at the valley bottoms a fluvial gravel occurs, and the scree slopes of the Slochd are coarse and unstable supporting little vegetation. The till is deeper on the lower slopes of Greenshiels Hill, possibly affording the opportunity for improving the vegetation type.

Figure 3:2 The geology of the farm

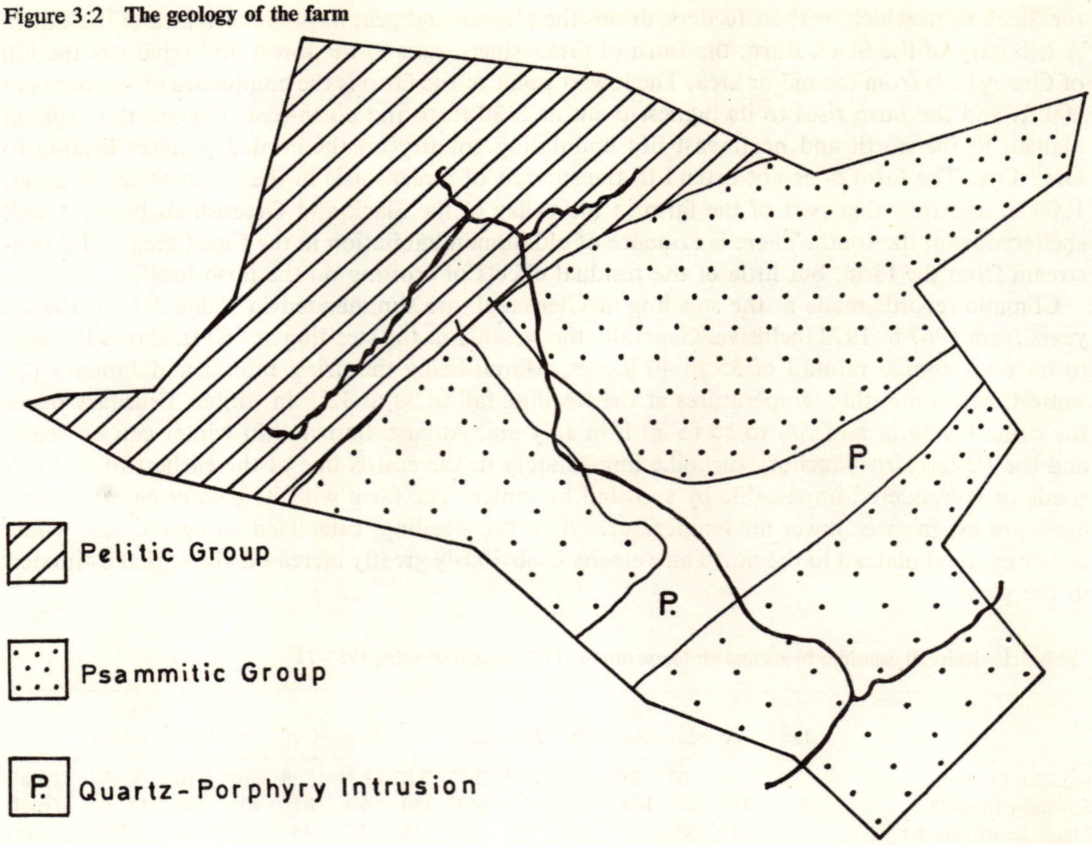

Vegetation

Figure 3:4, which is based on the ecological survey of Nicholson and Robertson (1958), summarizes the main plant association of the farm. This survey has been supplemented by a more detailed mapping of some of the paddocks of the farm with the object of using them as a base line from which to assess whether there is any ecological change induced by a deer stocking.

Figure 3:3 The soils of the farm

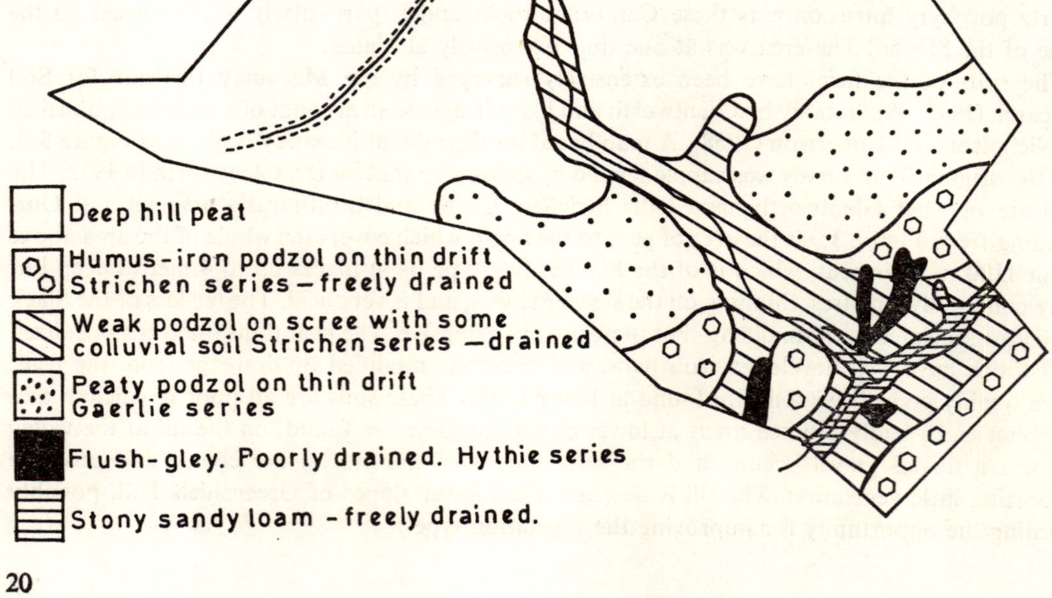

Figure 3:4 **The main plant associations of the farm**

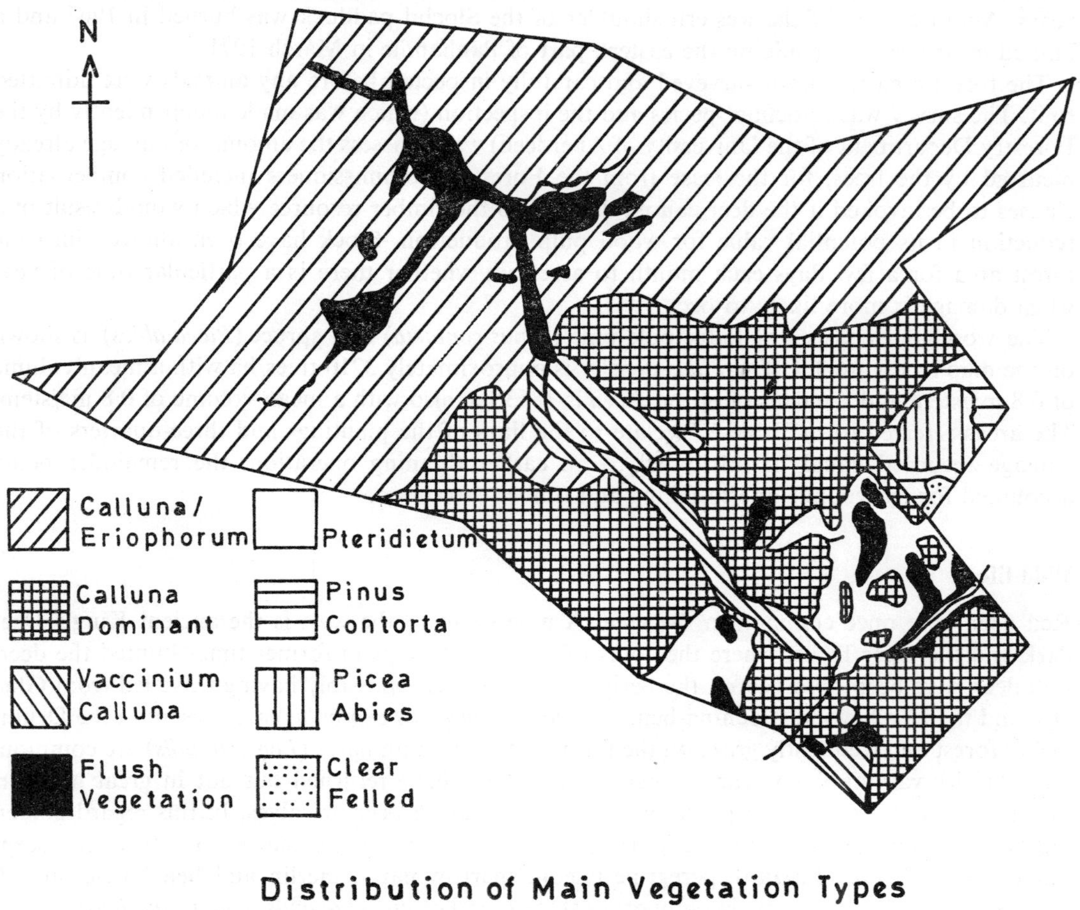

These separate surveys show that little change had taken place in the 15 years between Nicholson and Robertson's survey and the present time. Sheep stocking had been low during this interval, and the rabbit population, which certainly had earlier appeared to be a determinant of botanical change, had been slow to re-establish following its decimation by myxomatosis in the mid-1950s. The absence of change is perhaps remarkable.

Dominating the whole area are the plant associations containing ling (*Calluna vulgaris* Salisb.). In the north, on the wet conditions of the peat, *Calluna* is associated with sheathing cotton grass (*Eriophorum vaginatum* Linn.) and occasionally the common cotton grass (*E. angustifolium* Roth.). Indeed, the *Calluna* is co-dominant with the *Eriophorum*. Some cross-leaved heath (*Erica tetralix* Linn.) occurs together with sphagnum moss (*Sphagnum* spp.). As the blanket peat gives way to the podsol soils ling becomes completely dominant with small inclusions of bent (*Agrostis* spp.), tufted hair grass (*Deschampsia flexuosa* Beauv.) and woodrush (*Luzula campestris var. multiflora* Lej.). In turn, this *Calluna* domination gives way to an association with blaeberry (*Vaccinium myrtillus* Linn.), this being particularly apparent in the Slochd. On the lower slopes there is bracken (*Pteridium aquilinum* Linn.), varying in density and consequently associated with a variable number and density of other plants.

Commonly *Agrostis tenuis* is found where the bracken is least dense and *Deschampsia* and meadow soft grass (*Holcus mollis* Linn.) where the bracken is dense, tall and vigorous. Associated plants are the heaths *Calluna* and *Erica*, tormentil (*Potentilla erecta* Hampe.), bedstraw (*Galium* sp.) and sheep's sorrel (*Rumex acetosella* Linn.).

The peat, the valley of the Greenshiels burn and parts of the slopes are associated with wet flushes. Here the flora is dominated by rushes, notably the common rush (*Juncus effusus* Linn.) and the jointed rush (*Juncus acutiflorus* Hoffm.) together with sedges (*Carex* spp.) and grasses. In the limited areas of the valley bottoms there is a typical meadow association containing grasses (*Poa* spp., *Holcus* spp., *Festuca* spp. and *Anthoxanthum odoratum*) together with some clover (*Trifolium repens* Linn.).

Generally, the vegetation on the hill land can be stated to be typical of much similar land in the foothills of the Grampians. The state of the heather (ling) depends on the policy of muirburn practised. This had been recorded and from these records and observation it seems that some of the *Calluna* was as much as 20 years old but a large part had been burned within the previous ten years. About 5 acres of the western shoulder of the Slochd paddock was burned in 1967 and a limited muirburn was made on the eastern part of Panhandle in March 1971.

The forest area was both surveyed and carefully inspected before any animals were admitted to it. The survey was to define species and the inspection (which was made independently by the Forestry Department of the University of Aberdeen) was to assess the amount of damage already incurred by the trees, for the lease from the Forestry Commissioners included compensation clauses to be invoked if the deer caused damage to the timber resource which would result in a reduction in its potential value for wood pulp production. Stock have been allowed into the forest area for a few days each month to ascertain whether there is a particular time of year when damage is more likely to occur.

The woodland consists of lodge-pole pine (*Pinus contorta*) and spruce (*Picea abies*) as shown on the map in Figure 3:4. The *Pinus* density is approximately 3210 trees/ha with a mean volume of 0·8 m^3/stem and the density of the *Picea* is 1853/ha, also with a mean volume of 0·8 m^3/stem. The areas referred to in Table 3:2 relate to the lines of the plantings and three-quarters of the damage recorded could be attributed to an earlier thinning operation, the remainder being accounted for by wounds due to felling and timber extraction.

Wild life

Red deer were once common in the area. A mile or so to the west is the ancient King's Deer Park and the Deer Dyke, where the court of Scotland's kings in former times hunted the deer. Red deer have been observed on the periphery of the farm, possibly having travelled from Glen Dye and the hills beyond Cloch-na-ben. Roe deer (*Capreolus capreolus*) are present in the Drumtochty forest and frequently graze on the farm area. Mountain hares (*Lepus timidis*) are common, their density varying from year to year, while rabbits have returned but not in great number. Foxes occur but there are no predators which might affect deer or calves. In this regard golden eagles last nested on Slochd Crag in 1951 and since the farm was started one has been seen. The common kestrel, buzzard, peregrine falcon, sparrow hawk, merlin and hen harrier are all visitors to the area. The farm also supports a stock of red grouse (*Lagopus scoticus scoticus*).

Table 3:2 **Extent of damage to the lodge-pole pine and spruce in the forest area before admission of deer**

Area	Trees Examined	Trees Damaged	% Damage
Corridor through the lodge-pole pine	95	8	8
Transect bands 1, 4, 13 and 25			
(1) Lodge-pole pine	207	31	15
(2) Spruce	328	81	40

References

ANDERSON, J. G. C. (1945). The stratigraphical order of the Dalradian schists near the Highland Border in Angus and Kincardineshire. *Trans. geol. Soc. Glasg.* **20**, 223–237.

CAMPBELL, R. (1913). The geology of south-eastern Kincardineshire. *Trans. R. Soc. Edin.* **48**, 923–960.

MACAULAY INSTITUTE FOR SOIL RESEARCH (1947–48). Annual report, and soil map covering O.S. sheet 66/67 (Banchory/Stonehaven), Aberdeen.

NICHOLSON, I. A. and ROBERTSON, R. A. (1958). Some observations on the ecology of an upland grazing in north-east Scotland with special reference to Callunetum. *J. Ecol.* **46**, 239–270.

Chapter 4. **Design and effectiveness of fences and handling facilities**

The plans of the project called for considerable fencing. Fencing of the perimeter, of blocks of land for control of grazing and of smaller areas for hand-feeding was undertaken in the summer of 1970. At the same time a handling facility was built to enable animals to be restrained for scientific work. It soon became evident firstly that not all the fences had been ideally placed, and secondly that to aid the management of the stock and also the conduct of experiments some additional sub-division of the farm was necessary. This new work was completed in 1972. In addition, it was found that modification of the handling facilities was desirable. This was not so much because they had not proved satisfactory for handling deer, but because, at times, they provided conditions far from ideal for the people who were making measurements. It is possible to conduct a tuberculin test, to take blood samples and to weigh deer on a bad day in the depth of winter; the whole operation is, however, more efficient and less fraught with bad temper if a simple roof keeps off the sleet! Accordingly, the handling facility was roofed and a simple office/store erected to provide shelter. Thus, the present farm facilities have not only grown but have also been modified during the three years of operation. In the account of them that follows reference is made to these progressive developments only where these seem relevant to the design of commercial facilities.

A complication arises in describing the facilities precisely for during the period under review the unit of measure has been changed from the inch and foot to the centimetre and metre, while the value of pound sterling has also suffered a revision. While conversions of linear measurement are precise and unalterable with time, conversion of costs incurred in 1970, 1971, 1972, or indeed early in 1973 to those operating in the winter of 1973–74 are not. Dimensions of fences are all given in metric units. Costs are given in the original terms and have also been converted as far as possible into current values, taking cognisance of changes in costs of labour and materials.

Disposition of fences

Figure 4:1 shows the disposition of the fences, the names given to the various paddocks of the farm and their acreages. The central point of the farm is the handling area which occupies a bulldozed area at the end of the access road. Immediately adjacent is the feeding area. Access to the plateau of Birnie Hill and Goyle Hill is through the Panhandle. The fence between Goyle Hill and Panhandle was deliberately formed as a funnel to enable deer to be driven or 'drifted' down fence lines to a centrally disposed gate. While this arrangement worked well when deer were brought down into the V, when they were taken up it was impossible to move a group through the gate; they simply fanned out along the fences. To avoid this a 100 m net was placed in position and thus a small triangular area of about 1½ acres at the north-west of Panhandle could not be used by the deer. This area, together with a similar triangular area of 6 acres at the east of Birnie Hill, were fenced and trees are to be planted there.

Access to the Forest is through the holding-pen area. The wooded area itself has an internal fence parallel to the west fence of Panhandle, and these two fences form a ride which leads to the stag park. This arrangement was made so that stags could be brought from the stag park to the handling area without people being in direct contact with them, a matter of some importance at the time of the rut. The south-west perimeter fence of East and West Greenshiels was wrongly sited below the crest of the hill. As a result, after one moderate snowfall, drifting of snow completely covered it. Deer in Greenshiels at the time made no effort to escape.

Consideration of the map suggests that there might have been advantage in siting the handling facility centrally, though this would have entailed road construction. A central location might

Figure 4:1 Names given to the paddocks of the farm

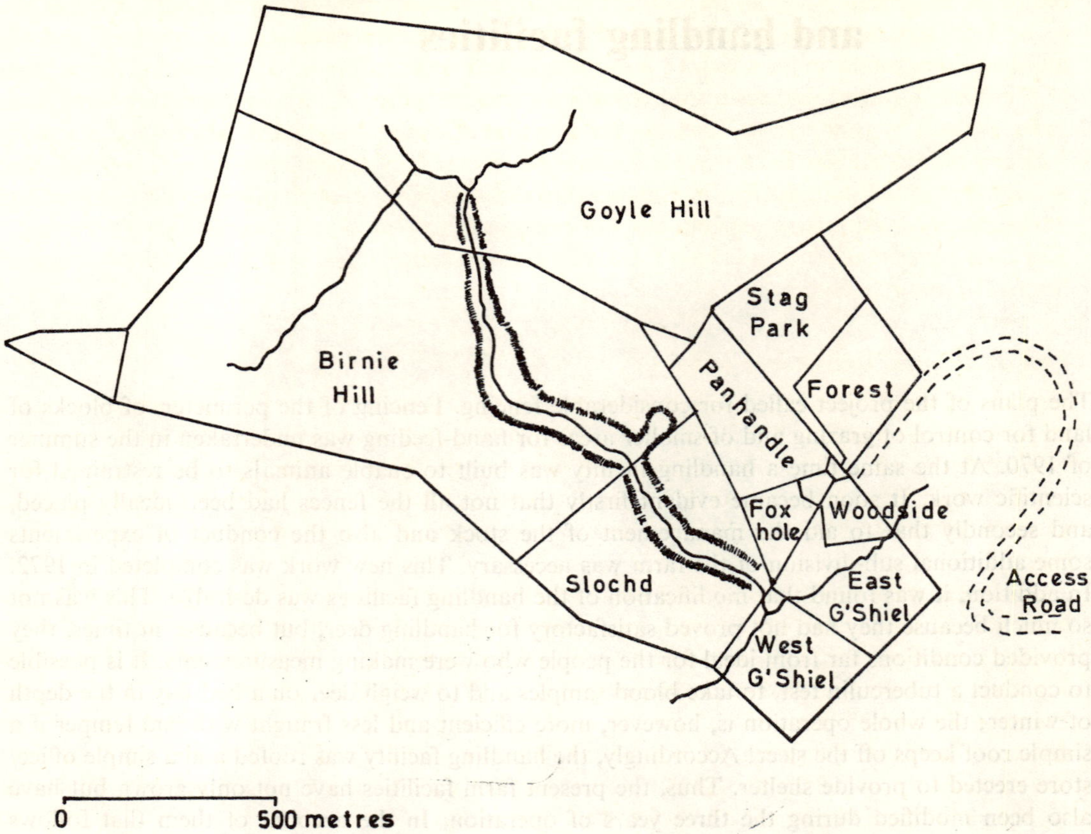

Figure 4:2 Location of the fences on the farm

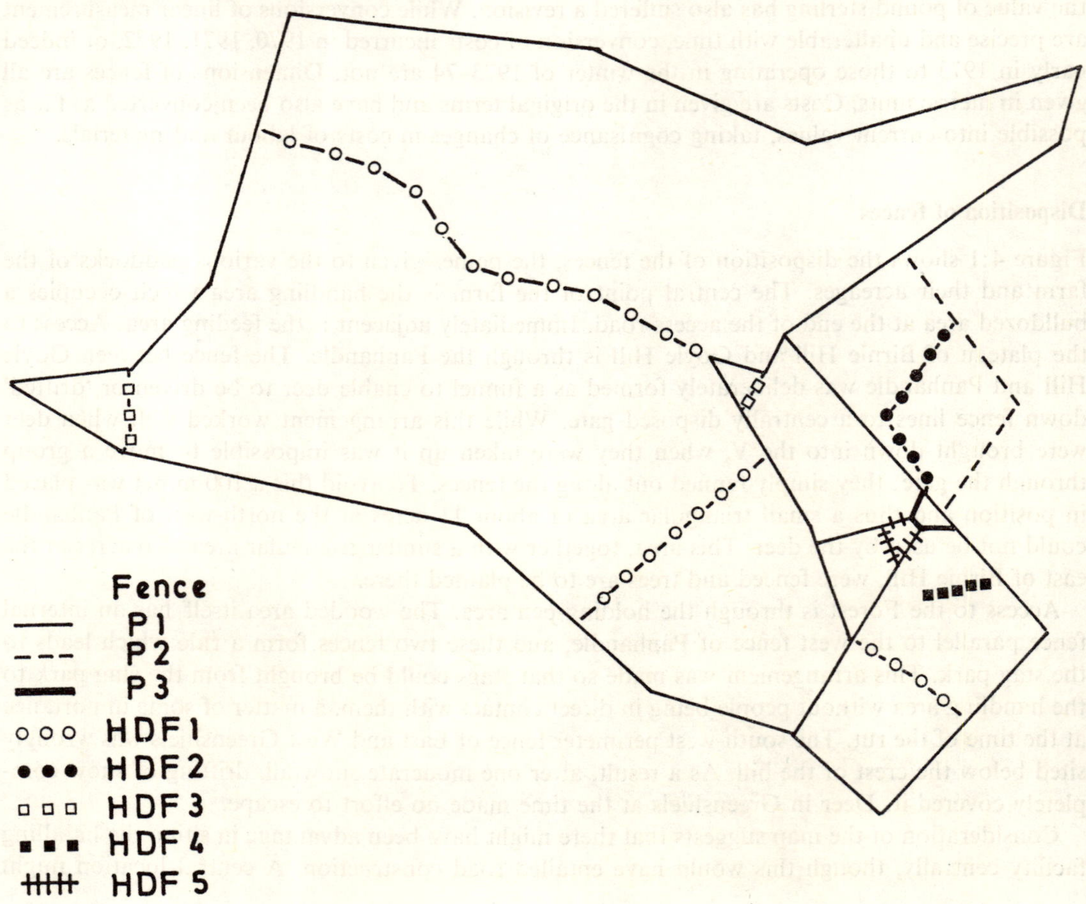

have enabled a radial distribution of paddocks to have been arranged. Certainly at the time of the rut, if, as appears likely under any deer farming situation, relatively small groups of hinds have to be kept separately, there is advantage in having a central paddock free of stock and also minimal distances through which to move stags. Reference to Figure 3:1, showing the contours of the farm, however, indicates that fence lines and the siting of handling facilities were in the main dictated by topography. Even so, the question of optimal lay-out of any future deer farms is clearly of importance. Even when the additional demands of experimental work are discounted, it does appear that minimizing the number of paddocks in an attempt to economize in fencing is undesirable. An estimate of current costs of erection of fencing on the farm is given in Table 4:1. The costs per mile of fence are those calculated from materials used, as given in Table 4:2, and the hours of labour required.

Construction and specification of fences

Fences on the farm can be divided into two main types; 'perimeter' type fences which are intended to be highly deer-proof and are all 7 ft (210 cm) tall, and 'internal' fences which, though they will contain stock, are not as tall being 6 ft (183 cm) in height or less. Two at least, and one in particular, of these internal fences are, in our opinion, sufficiently strong and serviceable to act as main perimeter fences. Figure 4:2 shows where the nine fences described below have been erected.

All fences were constructed using European larch or treated Scots pine timber and the wire used was 12 SWG high-tensile steel with a breaking strain of 100 tons/square inch. Table 4:2 specifies the dimensions of timber and accessories used in the construction of the fences while Figure 4:3 shows the profiles and dimensions of the 'perimeter' fences, P1, P2 and P3, and Figure 4:4 those of the 'internal' fences, HDF2, HDF3 and HDF5. The two remaining fences are not illustrated. Fence HDF1 is precisely the same as HDF2 except that the metal droppers are replaced by 5 × 3·5 cm droppers in timber. Fence HDF4 is intermediate between HDF1 and HDF3, and has a height of 5·5 ft (168 cm) with 10 strands of wire as in HDF3, but with the bottom strand on the ground.

Table 4:1 **Actual costs of erection of the fences and their estimated costs January 1974**

Fence	Erected in	Actual cost original specification	Estimated cost in January 1974* p/metre	£/mile
P1	Summer 1970	35p/yard	68·5	1102
P2	Spring 1972	60p/metre	92·0	1480
P3	Summer 1970	50p/yard	92·0	1480
HDF1	Summer 1972	48p/metre	67·25	1082
HDF2	Spring 1972	50p/metre	69·0	1110
HDF3	Spring 1973	40p/metre	55·75	897
HDF4	Spring 1973	36p/metre	61·50	989
HDF5	Spring 1971	55p/metre	75·50	1215

*Based on a contract for 1000 metres run of fence on hill ground including delivery of materials, distribution and erection.

Table 4:2 **Specification of timber and materials used in the construction of fences (cm)**

Component	Fences: P1, P2 & P3	HDF1, HDF2* HDF5	HDF3	HDF4
Strainers	320 × 15 T.D.	305 × 17 T.D.	270 × 15 T.D.	290 × 17 T.D.
Turning posts	320 × 13 T.D.	305 × 15 T.D.	270 × 12 T.D.	290 × 12 T.D.
Struts	350 × 10 T.D.	305 × 10 T.D.	300 × 10 T.D.	300 × 10 T.D.
Posts	275 × 9 × 9	245 × 9 × 9	215 × 9 × 9	230 × 9 × 9
Long droppers	218 × 5 × 3·5	190 × 5 × 3·5	158 × 5 × 3·5	172 × 5 × 3·5
Short droppers	102 × 5 × 3·5	—	—	—
Sheep netting	107, 16 swg 7·5 mesh	—	—	—

*Fence HDF2 employed steel droppers and arrowhead anchors.

Figure 4:3 **Profiles of the perimeter fences**

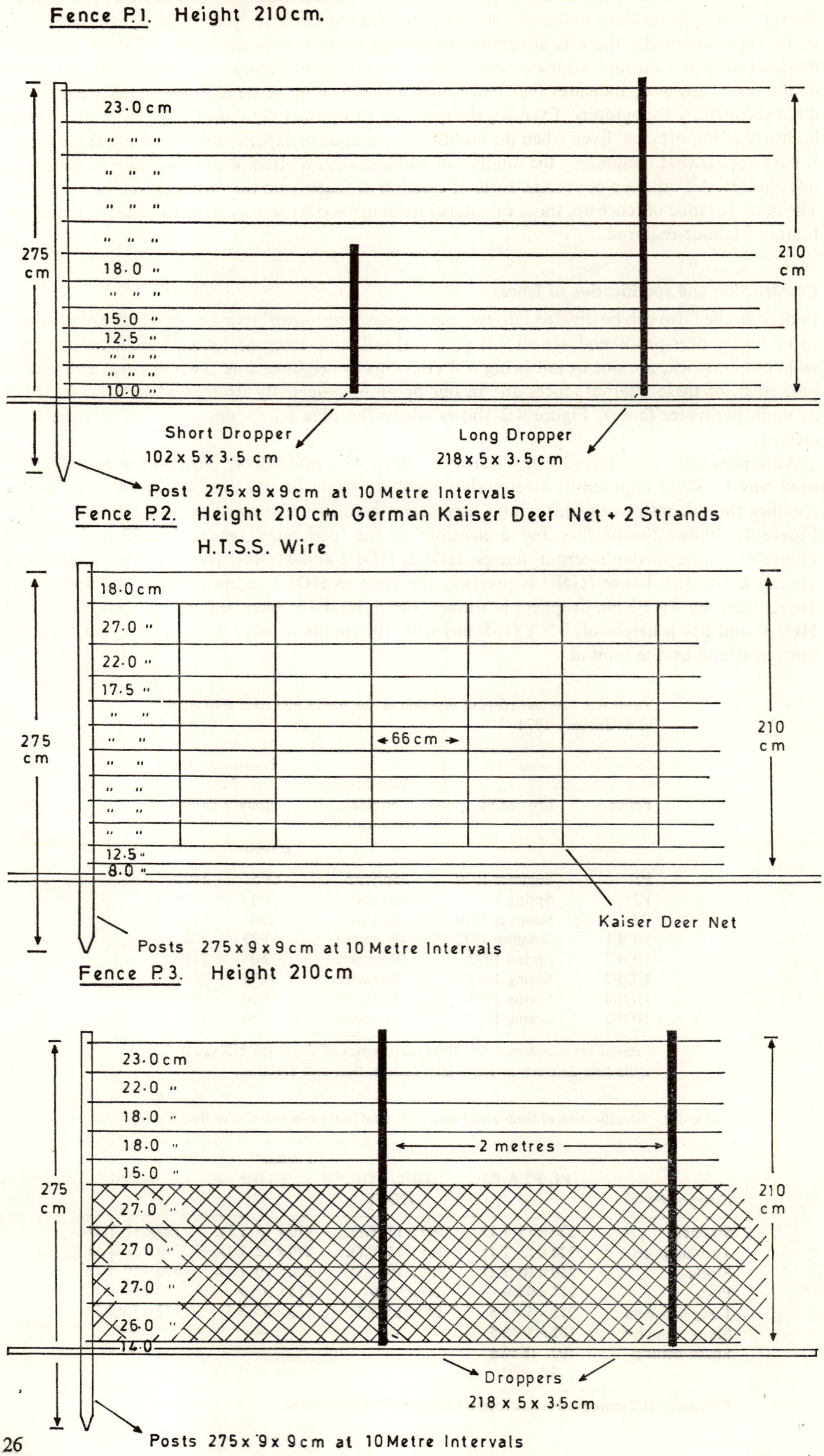

Figure 4:4 **Profiles of the internal fences**

Fence H.D.F. No. 2. 12 Strand H.T.S.S. Wire and H.T.S.S. Droppers

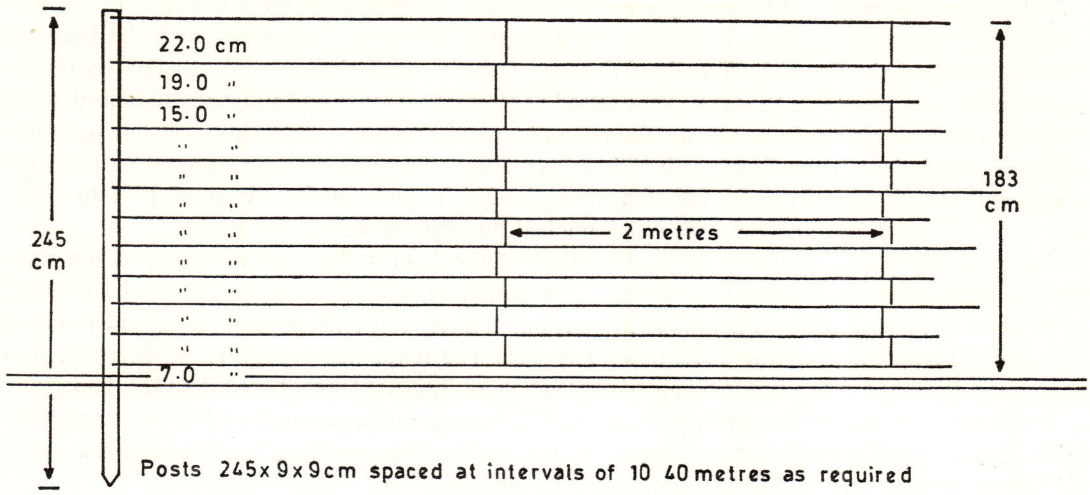

Fence H.D.F. No. 3. 10 Strand H.T.S.S. Wire-Wood or Steel Droppers

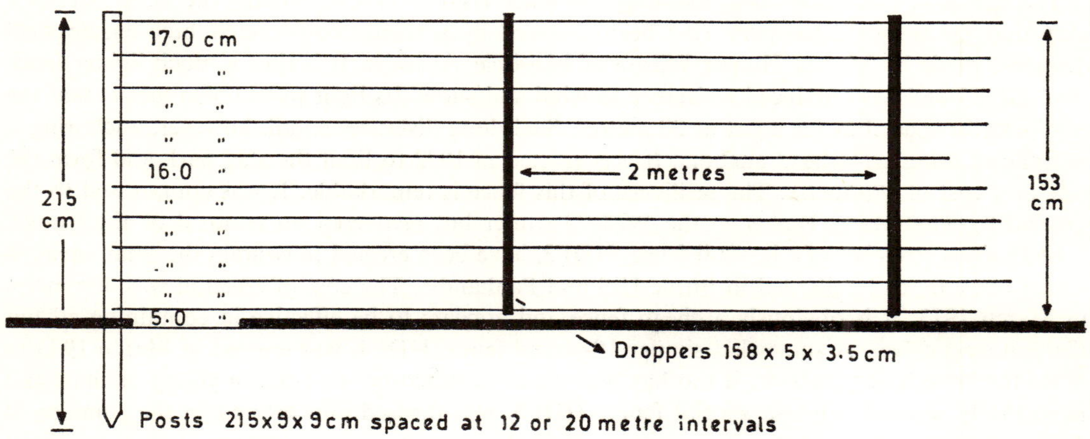

Fence H.D.F. No. 5. 8 Strand H.T.S.S. Wire with Rylock Netting

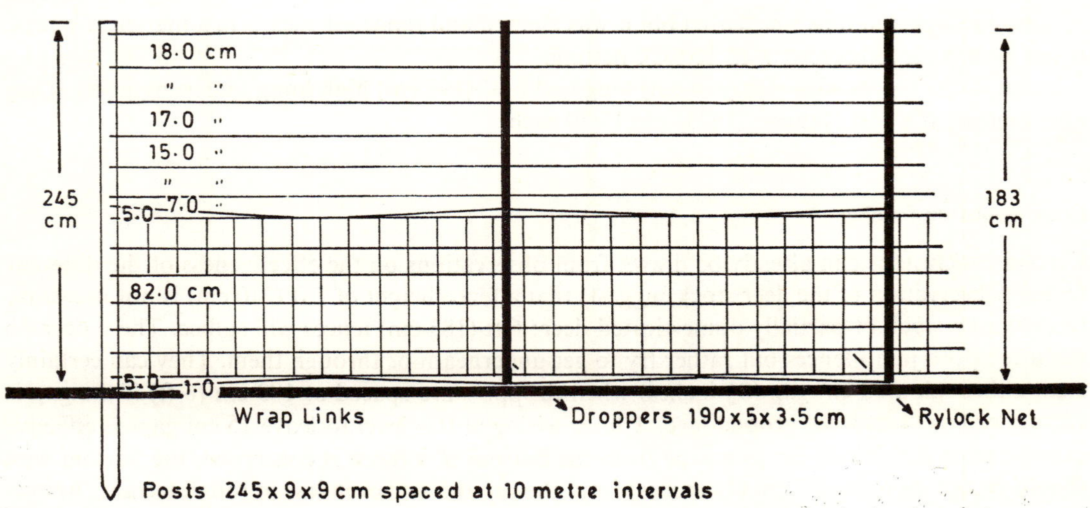

Effectiveness of fences

The experience gained about the effectiveness of the fences relates only to their ability to contain stock, how they have stood up to attrition by animals and their ability to withstand the weather when new. Nothing can be said about their useful life though perhaps a little can be inferred about their lasting qualities.

Perimeter fence no. P1 is the main perimeter fence and was erected in the summer of 1970 under contract at a cost of 35p/yard. Its design was arrived at following discussions with the Forestry Commission and the Scottish Woodland Owners Association. It has proved satisfactory for no animals have escaped from the farm. When used as an internal fence and subjected to considerable pressure due to stock movement—as, for example, on the borders of Foxhole—it has been less effective. Hinds have gone through the 23 cm gap above the short dropper, suggesting too wide a gap and too short a dropper. This fence was buried in snow in the winter of 1971 to 1972; staples were pulled out and the wire was stretched but repairs were easily made.

Perimeter fence no. P2 is used on the boundary of the farm in the wooded area and is a commercial fence made by Kaiser of Germany and available in rolls of 100 m or 200 m with a height of 183 cm. It was erected to a height of 210 cm but is likely to be effective at its designed height. The cost of erection in spring 1972 was 60p/metre. The fence has proved to be effective, it is easily and quickly erected but it seems likely that as an internal fence it would suffer damage.

Perimeter fence no. P3 was erected in the handling areas in the summer of 1970 and cost 50p/yard. A strong calf-proof fence was required here and the sheep netting has been effective. Such netting must be at least 3·5 ft (107 cm) and of good quality for this purpose.

Of the so-called internal fences, fence HDF1 has proved extremely effective with all classes of stock. It was designed on the basis of experience with fence P2 as an internal fence and was used to divide Goyle Hill from Birnie Hill, to cut off Slochd and divide Greenshiels Hill into two. It was erected in the summer of 1972 at a cost of 48p/metre.

The design of this fence led to the design of fence HDF2. This is virtually the same as HDF1 save that the wooden droppers have been replaced by specially constructed high-tensile steel droppers made by Messrs. Hunter Wilson of Maybole, Ayrshire. It is thus a much lighter fence than HDF1 and has a reduced resistance to wind and snow. Its light weight also means that the posts can be spaced as far apart as 30 metres. Such long distances entail, however, anchoring it in hollows using arrowhead anchors. It was erected in 1972 to limit the stag park and form the ride at a cost of 50p/metre. The resilience of this fence is remarkable. It was compressed to the ground by wind-blown timber in the 1972–73 winter but recovered on removal of the trunks.

Only short stretches of a less tall fence, HDF3, have been erected to contain the small areas in which trees are to be planted in Birnie Hill and Panhandle. The cost of erection was 40p/metre in March 1973. It is obviously a cheap fence and appears to be effective in containing all but the youngest calves. A somewhat similar 10-strand fence, HDF4, was erected in March 1973 to fence the Woodside paddock; it too has been quite satisfactory. To contain young animals and particularly weaned calves a special fence, HDF5, was devised and erected in the summer of 1971 at the cost of 55p/metre. This consists of 'Rylock' netting hung between plain wires about 5 cm above and below it. This fence has proved very stock-proof and robust.

Other fencing materials have been examined. One consisted of a polythene slack twist, square mesh, net corded at its edges which was hung on three wires stretched between posts spaced at 10 metres. Although the net was provided free of charge it cost 20p/metre to erect. No animals became entangled in it as was feared but it was chewed and rendered useless in a matter of weeks. It has now been reinforced with Rylock netting.

Gates in the fences were 9 ft (274 cm) long and 6′6″ (198 cm) high hung on strong posts. They were erected at a cost (January 1974) of £17·50 each.

Conclusions on fencing

Certain conclusions can already be drawn from observations on the effectiveness of the different fences. Observation of the deer stock suggests that a fence height of 7 ft (210 cm) is not necessary to contain a stock of partially domesticated deer; 6 ft (183 cm) fences will suffice. These deer do not attempt to jump fences but rather try to get underneath or through them. They can certainly escape through a 23 cm gap (8½ inches) when droppers are spaced at 183 cm (6 ft) if the gap is 122 cm or less (4 ft) from ground level. Closer wiring at this level to leave 15 cm gaps is effective in preventing escape. As far as escape from the bottom of a fence is concerned, the bottom wire should be not more than 7 cm (2¾ inches) from the ground and even less in rough terrain. Arrowhead anchors to pull the fence down into hollows are superior to short posts. Shorter fences have the great advantage that they offer less resistance to wind and snow, and can, therefore, be made lighter than taller ones. Droppers of steel rather than of timber also reduce wind resistance and fence HDF2 thus has much to recommend it. Whether shorter fences than the present 5 ft (153 cm) ones recently erected will be sufficiently effective to act as internal fences on a deer farm will be explored; they would certainly reduce costs. Table 4:1 summarizes the costs of the fences in their original terms and corrected to prices ruling in January 1974. Though not classed as perimeter fences, fences HDF1 and 2, and more particularly HDF2, could serve this purpose, while fence HDF3 should be effective as an internal fence.

It is of interest to compute minimal fencing requirements for deer farms of different sizes. If the maximal area of grazing blocks is taken as 160 acres—$\frac{1}{4}$ square mile—then as the size of the deer farms increases from 1 square mile (640 acres) to 4 square miles, although the length of perimeter fence/square mile decreases, that of internal fencing increases and a farm of any size is likely to require about 6 miles fence/square mile of farm simply to provide blocks of land for grazing. If stocking density is about one hind to 6 acres it is possible that this amount of fencing together with some additional fencing of stag enclosures and feeding areas would suffice. If, however, density is higher than one hind to 6 acres then the sizes of separately pastured groups of hinds at the rut would exceed 25. From the preliminary information given in Chapter 8 this might not result in satisfactory conception rates. Additional fencing would thus be necessary. From such calculations it appears that deer farming requires a minimal investment in about 7 miles of fencing/square mile whatever the size of the farm. The cost falls slightly with increased size of farm but only because the ratio of the more expensive perimeter and special fencing to the less expensive fencing of internal divisions falls with size.

General plan of the handling facilities

The handling facilities of the farm consist of the smaller paddock areas, the pennage for animals, boxes for special purposes, the crush and race, the weighing pen, the shedder, and the feed stores. The relation between the paddocks—Foxhole, Woodside and the feeding area—and the system of pennage is shown in Figure 4:1, Figure 4:5 shows the disposition of the pennage and Figure 4:6 the working area for handling the animals directly.

The paddocks are merely fenced areas of the farm, small in size and used to hold groups of deer for periods of 24 h or more. They have proved very useful as large collecting areas for the stock in which they can be held before detailed examination on a following day. In addition, the feeding area acts as a confinement area for younger animals and a place in which they can be fed.

Figure 4:5 **Disposition of the main pennage and handling area**

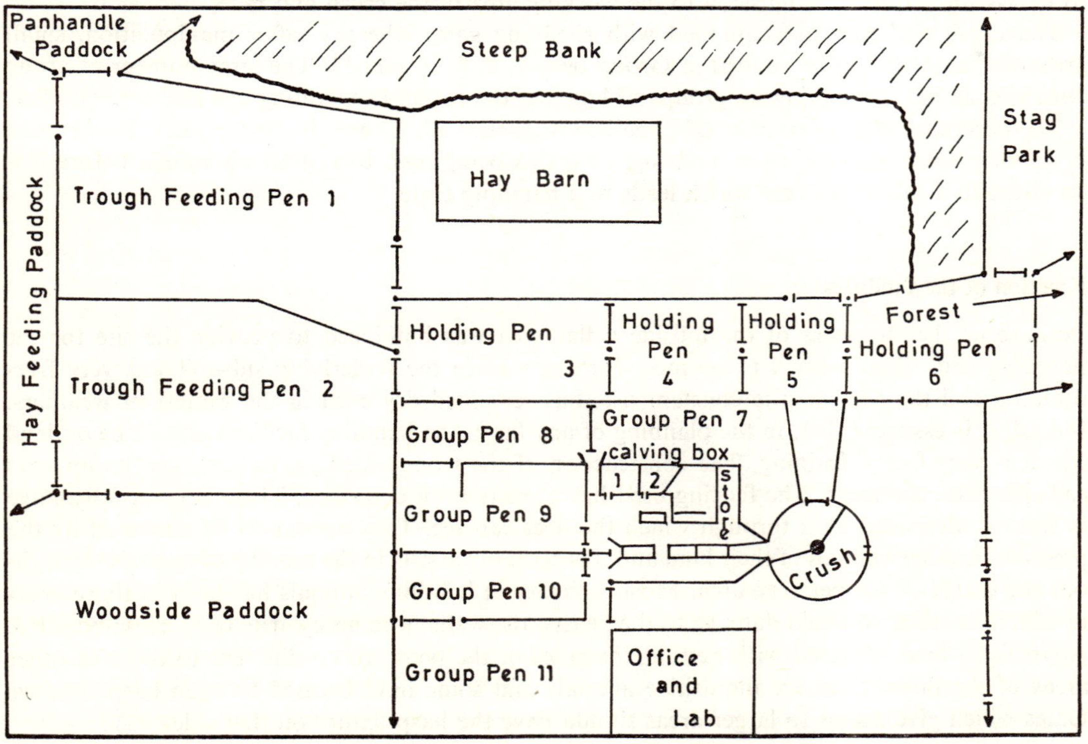

When weaned calves were first brought to the farm they were initially confined to the 1 acre of the feeding area with a view to establishing this as their 'home area' before they were released to the larger acreages of Woodside and Foxhole. In addition, it was probably only in this way that the weaned calves could be encouraged to eat. The same procedure was used with the weaned calves from the 1973 calving.

Figure 4:6 **Plan of the working area of the handling facility**

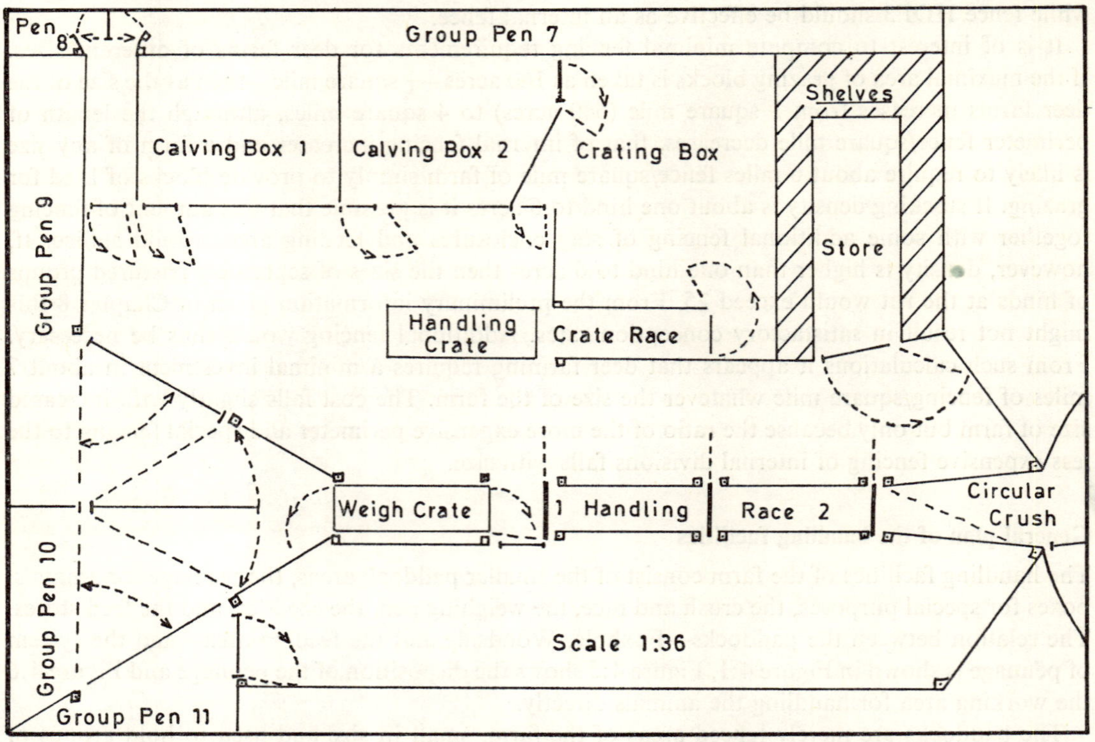

As shown in Figure 4:5, the pens were of different size. The large trough-feeding pens 1 and 2 were designed for holding groups of animals immediately before their access to holding pens 3, 4, 5 and 6 which together, when their gates are opened, provide a continuous 15 ft access way to the forest area and the stag park. Holding pens 3, 4, 5 and 6 also enable small groups of deer to be assembled before admission to the working area of the crush and race.

The crush and race are equipped with shedding gates whereby, after manipulation, small groups of animals can be penned in Group pens 7, 8, 9, 10 and 11. The large number of group pens is essential for making up groups of hinds at the rut and separating age and other classes.

The confinement area consists of three boxes labelled 12, 13 and 14 on the plan. Two consist of calving boxes and the other a crating box, this being used to restrict an animal before it is admitted to a subsidiary race which leads to a handling crate.

Erection of the facilities

Because of the steepness of the hillside a flat area was bulldozed to provide the site for the handling pens. This proved to be most fortunate since the underlying sub-soil was very free-draining and the site, despite constant use, has remained dry even in the wettest of weathers. Indeed, it is essential that, in the planning of any farm, the handling facilities should be on land which is very freely draining. The construction of the trough-feeding, holding and group pens calls for little comment. The fencing used has already been described. There is great advantage in having wire-sided pens through which the deer can see. They appear to be reassured by the presence of other animals of their kind and this certainly assists in the moving of animals throughout the whole of the pennage area. Most difficulty in handling animals has been in those areas in which, in order to avoid damage to the beasts, the walls have necessarily to be close-boarded. Apart from being covered with netting, the gates of the pens are no different to those in other areas of the farm. Deer are inquisitive animals and some have learned to open hasp fixtures. Gates which give access to larger areas should have the hasp fixture on that side.

The whole of the working area was concreted, four inches of a coarse mix being found adequate, and was finished with a wooden float. The main corner posts, which were set in the concrete, were second-hand telegraph poles and these hold the roof. Subsidiary posts were of 4 × 4 European larch, also set in the concrete. The circular crush is 18 ft (5·5 m) in diameter draining to the outside. It consists of galvanized iron built in sections to a height of 7 ft (213 cm). A central steel pillar carries two wooden free-moving gates, supported on wheels, which are used to manipulate animals towards the handling race and also to a door which is at such a height that it acts as an off-loading bay. The main door admitting animals to the crush is also 7 ft (213 cm) high and is close-boarded on both sides. The handling race is 17 inches (42 cm) wide and consists of two units each 4 ft 6 in long (137 cm). These are separated by a wire-mesh sliding door and once one animal has been admitted to the system others follow it. The two units act, one as an operational area and the other as a decoy, though in fact, both are usually used. The race is close-boarded throughout and each of the separate units is equipped with inspection ports measuring 116 × 47 cm on each side. These ports allow animals to be handled and are at such a height that if opened with care animals do not immediately try to escape through them. They are used in particular for identification of animals since they readily allow the reading of ear tags. A gap between unit 2 and the weighing crate, closed on one side by a door, allows animals to be taken to the group pens without being weighed.

The weighing crate is shown in Figure 4:7. It consists of a welded frame sheeted with 10 gauge 2 in (5 cm) mesh welded wire. This is suspended from a Salter No. 199T crane weigher which has a 20 in diameter dial, capacity 300 kg with 1 kg divisions and readable to 0·5 of kg. The frame measures 4 ft 6 in long and is 6 ft high and 18 in wide (137 × 183 cm), and has doors at each end. The cage is stabilized top and bottom by sleeve dampers and its floor of 10 gauge woven wire ¾ in (1·9 cm) mesh is about 6 in above the concrete floor of the race. Deer have no difficulty in entering it. The dial of the balance is 8 ft 6 in (168 cm) above ground level and a tractor seat was

Figure 4:7a **The weighing cage front elevation**

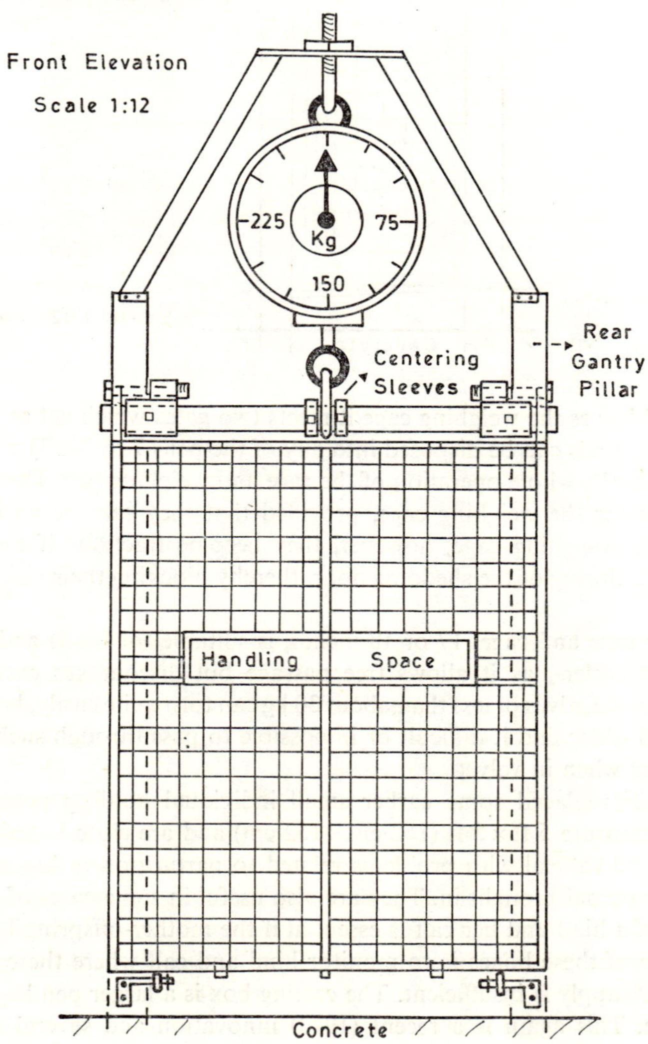

erected so that weights could be read easily. The balance has proved very robust and reliable and weighing can be undertaken quite rapidly. The weighing cage and balance were erected at a price of £250 in 1970; a tender for their erection at that time was £800.

Figure 4:7b **The weighing cage side elevation**

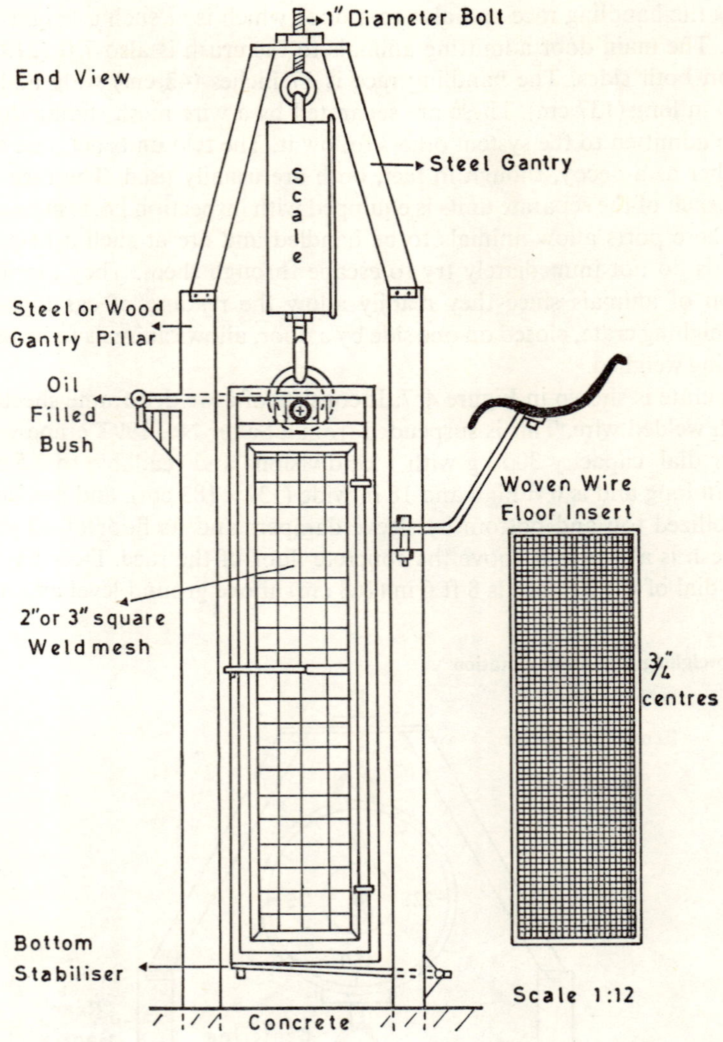

When the animal leaves the weighing cage it meets two gates which act as a shedding arrangement whereby the animals can be disposed into any of the pens 7 to 11. The shedding gates have proved very useful in the whole operation of the race and weighing pen. Deer will come forward into the race and enter the weighing cage, provided that they can see an exit and other deer beyond. Once in the weighing cage, however, they become fractious if they can still see this escape route and the closing of the shedding door, thereby blocking their vision, ensures that they settle quickly.

The width of the race and cage, 17 or 18 inches, is suitable for hinds and for young stags or older ones without antlers, as it allows free passage but discourages excessive struggling or attempts to turn round. Calves of less than about 30 kg can turn quite easily, however, and antlered stags of 2 years and older find it difficult or impossible to pass through such a width and easily damage their antlers when in velvet.

The calving boxes replaced some earlier small individual holding pens which proved unsatisfactory. They measure 5 ft × 5 ft (152 cm × 152 cm) and are close boarded to 7 ft (213 cm). They were introduced to deal with problems related to parturition in hinds when these can be recognized and the animal brought in. They are also useful in some cases of mismothering when the close penning of a hind and her calf is essential if the mother-offspring bond is to be forged. A further use made of these boxes is to monitor hind and calf where there is reason to believe that the hind's milk supply is insufficient. The crating box is a larger pen leading to a small race and handling crate. This again is a recent (1973) innovation and several designs of handling

Figure 4:7c **The weighing cage plan and view of support system**

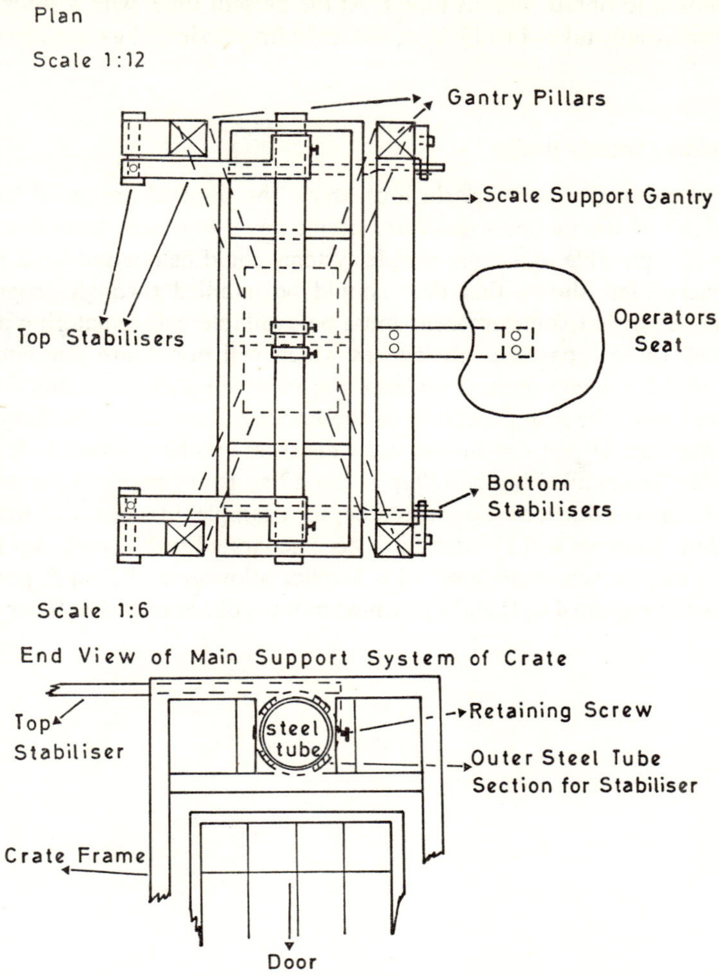

crate have been tested. None are yet entirely satisfactory and they are not described here. However, for the very close examination of deer a method of greater physical restraint than the simple confinement in a race is necessary if the safety of those handling the animal is to be assured.

The storage facilities for feed consist of a roofed store within the working area which holds concentrated feed and veterinary requisites. In addition, a hay-barn is to be erected to hold winter fodder. This will be of conventional construction. Space is to be kept vacant adjacent to the hay-barn to facilitate expansion.

Operation of the handling facilities

The whole farm stock is routinely gathered once a month when all are passed through the working area to be individually identified and weighed. On these occasions samples are usually taken for scientific purposes and any tests—such as tuberculin tests—made. The handling facilities are also used on occasions other than these to deal with smaller groups of animals. At monthly gatherings the stock is collected the previous evening, held in the holding paddocks overnight and sufficient animals to be passed through the system the following morning are brought to pens 1 and 2 at about 8 to 9 a.m. They are taken through holding pens 3, 4 and 5 in groups of about 30 and then are moved into the crush. Holding pen 6 is useful since it enables animals to be held and separated without passing through the working area. Experience soon showed that young calves must be separated from the hinds as soon as animals are fairly closely penned, indeed all age groups should be separated. Once a group of animals unlikely to fight has been assembled in pen 5 they are moved into the crush. One leaf provides a boundary wall with the line of the race and the other leaf is moved to accommodate the 30 or so animals admitted. This second leaf is moved up as the number of animals diminishes. In the race they are identified, moved into the weighing pen and thence to the group pens. When the whole operation simply entails identification and weighing it is possible to run one animal through in 1 to 2 min. On most occasions, however, a number of additional operations take place such as the collection of faecal samples from the rectum, the taking of hair samples, examination of teeth or the physical restraint of the animal to obtain venous blood. At the present time, with a stock of 150 animals, the whole operation usually takes 1 to 1½ days, not including the initial gathering or the subsequent dispersal.

Conclusions on holding facility design

As with fencing, some preliminary conclusions about the optimal design of handling facilities can be drawn. Many of the facilities described above have, however, been designed for experimental work and it is possible that more simple systems could be devised for a fully commercial enterprise. Experience has shown that deer should be handled through progressively smaller paddocks and pens and that sufficient pens must be available before entering a crush and race to enable the stock to be separated into classes which will not injure one another. Indeed, it could be argued that for a commercial undertaking relatively little economy could be made in the number of pens with the sole exception of the handling box and subsidiary race. From the behaviour of animals under pen conditions stock densities can be estimated. Where animals are being held for a few hours awaiting handling about 12 sq ft are required per adult beast. Once the animals are closely confined, as they would be, for example, in a crush or initial holding pen, less space is needed. Animals will by then have been separated into age classes and calves up to six months of age can be accommodated with a space allowance of 3 sq ft per animal. Calves six to nine months old require 4 sq ft and those nine months old or over 5 sq ft per animal.

Chapter 5. **The capture and rearing of calves**

In Scotland most red deer calve in early June. A few calve in the last week of May and by the end of June nearly all the pregnant hinds in a herd have given birth. Some calves may be born late—in July and even in August; these seldom survive the succeeding winter. Calving in the west of Scotland is earlier by about a week than in the central Highland and the eastern Grampian regions. Calving in the wild is thus compressed into a remarkably short period of time.

The long experience of the Red Deer Commission and the Nature Conservancy has shown that, in the wild, as the time for calving approaches a hind seeks a place on which to give birth which is apart from the rest of the hind herd. Unlike the hill ewe the area she selects is not necessarily the same one in which she herself was born. Hind herds inhabit defined geographical areas beyond which only very unusual climatic conditions, population pressure or emergencies will drive them, but their choice of calving area seems to be dependent more on the circumstances at the onset of the first stage of labour than on a hefting to a specific territory. The three stages of labour may last for as long as 24 h and during most of that time the hind remains separated from the rest of the herd. She may, however, rejoin the herd for grazing if the first stage of labour is intermittent. After giving birth the hind with its calf begins to move towards the herd, but the calf soon lies down. Then, if the distance is not great, the hind may rejoin the herd but, in general, she hovers not very far away from the calf's resting place for the first two or three days, returning from time to time to suckle. An account of the calving behaviour of hinds under farm conditions, which may be compared with this general description of wild animals, is given in Chapter 7.

Capture of calves was undertaken in June of 1970, 1971 and 1972. The methods used in 1970 and 1971 are described here very briefly since the capture was for the most part undertaken by stalkers and members of the staff of the Red Deer Commission and the Nature Conservancy. In 1972, however, capture was undertaken by teams from the Institute and Organisation.

In 1970, an attempt was made to obtain 30 calves for the farm, it then being envisaged that a build-up of stock over two years would suffice. Despite generous help from the Red Deer Commission and the Nature Conservancy and from private sources a late start was made and only nine animals were obtained for the farm, a further 12 being retained at the Institute to be used to examine aspects of the nutrition of the reared calf and thus anticipate some of the problems that would be later encountered on the farm. This was clearly not an auspicious beginning and so in the early spring of 1971 approaches were made to a number of landowners in the north of Scotland to ask whether they would collaborate in a major effort to stock the farm. They, and their stalkers, agreed with enthusiasm and an arrangement was made with the kind help of the North of Scotland College of Agriculture centre at Inverness to provide a staging post at which calves could be held overnight on their long journey from the north to the east of Scotland and the rearing quarters at the Rowett Institute and at the Glensaugh farm of the Hill Farming Research Organisation. In addition, a small team was sent to Rhum to catch calves for again the Nature Conservancy had agreed to help. The result was that 75 calves were obtained. There were, however, problems. Despite the generous provision by the North of Scotland College of Agriculture of a staging post for feeding and rest in Inverness, three calves were dead or very weak on arrival at the Institute. By contrast there were no casualties in a group of 11 calves which had had a long sea and road journey from the Inner Hebridean Isle of Rhum. These calves had been gradually accumulated in a steading on the island and kept there until all were trained to feed from a bottle before being transported. It was therefore decided that the 1972 intake should also be assembled in groups of 10 to 20, with a local feeding centre for each group. Accordingly, and with the much-appreciated co-operation of landowners on Islay and on Jura, and of the

Nature Conservancy on Rhum, a further capture of deer was made in 1972. The methods used are described below; the catching methods themselves are virtually the same as those employed by the staff of the Red Deer Commission in their annual marking of calves, undertaken to establish the basic parameters of the population dynamics of wild deer.

Capture of deer calves—1972 experience

Preparations for assembling a group of calves required the setting up of a convenient temporary headquarters with living accommodation for the catchers and a dry barn for the calves. Pens were made with wooden hurdles and the floor was covered with deep clean straw.

Catchers worked in teams of two and set out each morning with the ambition of bringing back two calves per catcher in a day. Each was therefore equipped with a haversack for securing one calf, and a hessian sack and string for securing a second. Other essential equipment was a pair of binoculars (or a telescope) for each person and one catching net for each pair. The net, used for pinning a lying calf to the ground, was made of nylon and was secured to a metal rim 70 cm in diameter to form a hemisphere. The metal ring with the net was fixed to the end of a light-weight pole 3 m long.

Early in the morning the two catchers climbed high from a valley in which hinds were likely to graze and down-wind from them. A suitable vantage point was sought and the herd watched for hinds with young calves at foot. Continual scanning was necessary because the appearance of a calf beside its mother happened unobtrusively. The catchers then had to communicate with each other and both focus on the same hind and calf. The usual sequence of events was that the calf sucked for a few minutes and then had a play-period of about 10 minutes in which it gambolled in the neighbourhood of its dam. The calf then left its dam abruptly and sought somewhere to sleep. Some calves travelled up to 500 m before lying and even then, after lying for a short time rose again to seek another bed. Meanwhile the watchers had to discuss and agree upon the final resting place of that particular calf before turning their attention back to the herd to look for more. They also looked for hinds lying apart from the herd which might show signs of labour or the movement of a new-born calf alongside.

After a few hours, when the location of one or more calves was known, one catcher descended and the other directed him by signals to the first calf to have been located. In descending he deliberately disturbed the hinds who then retreated, leaving their calves behind. The final approach to the calf entailed a stealthy approach, a confirmatory signal from the look-out on the hill to indicate that the catcher was 'on line' and a final pounce to pin the calf to the ground. The calf was then put into the haversack with its head protruding and the haversack secured with its draw-string. Other calves were similarly secured by the two catchers working as a team until a maximum of four calves had been captured. All were then back-packed out of the valley to the origin of that day's expedition, whence the calves were driven by motor vehicle to the feeding centre.

At the feeding centre each calf was weighed, identified with a numbered chicken wing band in each ear and offered warm re-constituted ewe milk substitute from a bottle through an artificial lamb teat. Newly born calves learned very quickly to drink from a bottle and very quickly adapted to living in groups in the straw-bedded pens. They soon greeted the arrival of a handler with bottles of milk with obvious enthusiasm and approached voluntarily to suck from the bottle. Older calves were more wary and stubborn and had to be sustained during the first 24 h and occasionally for two or more days by gently forcing them to swallow the milk substitute. They gradually adapted to communal living. The calves were fed four-hourly by day with an eight-hour overnight interval.

Preparation for transport to the deer farm consisted of feeding them once during the night and again in the early morning after which each calf was sewn into a hessian sack with its head protruding. The calves were then carefully placed in a motor vehicle for transfer by car ferry and by road. A roadside feed of warm re-constituted milk was given on the way, the calves being taken out of the vehicle one at a time for feeding to ensure that none was missed. A further feed was ordered by telephone to be ready at the estimated time of arrival at the deer farm.

Weights of calves at birth

Weights of calves were recorded soon after capture and mean values are given in Table 5:1 for both stags and hinds in each of the three years. Apart from showing a difference in weight as between stags and hinds it is difficult to analyze these data further. The precise age at capture of the calf is not known nor is the age of the dam and no estimate can be made of her nutritional state at calving. All these factors might affect birth weight. An analysis which is more meaningful relates to factors affecting the birth weight of calves born on the farm in the summers of 1972 and 1973, that is of the 52 animals for which precise information is available. These data can be used to indicate the effect of sex of calf and age of dam on birth weight as well as any effects of season of calving and the time of birth. Table 5:2 summarizes these data.

Table 5:1 **Weights of calves at capture in 1970, 1971 and 1972**

	Stags		Hinds	
Year	No. captured	Weight (kg)	No. captured	Weight (kg)
1970	5	6·9	4	6·6
1971	20	7·8	55	7·5
1972	19	7·5	35	7·8

Table 5:2 **Birth weights of calves born on the farm in 1972 and 1973**

		Stag calves		Hind calves	
		No.	Weight (kg)	No.	Weight (kg)
1972	1st calvers	3	7·67	2	6·75
	2nd calvers		—		—
1973	1st calvers	18	6·82	23	6·28
	2nd calvers	4	8·12	2	6·80

These data showed that on average stags weighed $0{\cdot}65 \pm 0{\cdot}21$ kg more than hinds, that second-calvers produced calves weighing $1{\cdot}02 \pm 0{\cdot}33$ kg more than first-calvers and that there was a real difference between the two years of $0{\cdot}68 \pm 0{\cdot}37$ kg in favour of 1972. In addition, the later a calf was born the heavier it was; calf weight at birth increased by 0·33 kg if it was born a month late. There was a broad relationship, irrespective of age and calving date between the weight of the hind at conception and the weight of her stag or hind calf. This is illustrated in Figure 5:1. It seems probable that the difference in birth weight between first and second calvers is entirely accounted for by this difference in body weight at conception.

It is thus understandable that the weights of the calves captured in the wild at ages presumed to be between one and five days varied appreciably. It is difficult to state whether there were any regional differences in birth weight in view of these other causes of variation.

It is of interest that these late born calves did not catch up entirely in their growth by September. Generally speaking, in September a calf which was born on July 1st will weigh 7·0 kg less than one born on June 1st. Rate of growth in the first six weeks of life was greater for animals with the heavier birth weight, every additional kg at birth being associated with an additional $2{\cdot}1 \pm 0{\cdot}2$ kg of weight at 42 days of age.

Artificial rearing

Table 5:3 shows the body weights of the surviving captured calves during the 8 week period when they were fed milk substitute and weaned on to solid food, their weights in October when they had been pastured at the farm for a few weeks and also those recorded in December when they were about 6 months of age. Similar data for the home-bred naturally suckled calves of 1973 are also given. There were clearly wide differences between groups in different years in the early growth attained but by December these differences were much smaller.

In 1970 the deer were reared at the Rowett Institute on a proprietary cow-milk substitute made up as for bovine calves and were weaned on to artificially dried grass and a home-made

Figure 5:1 **The relation between the weight of the hind at breeding time and the weight of her calf when born**

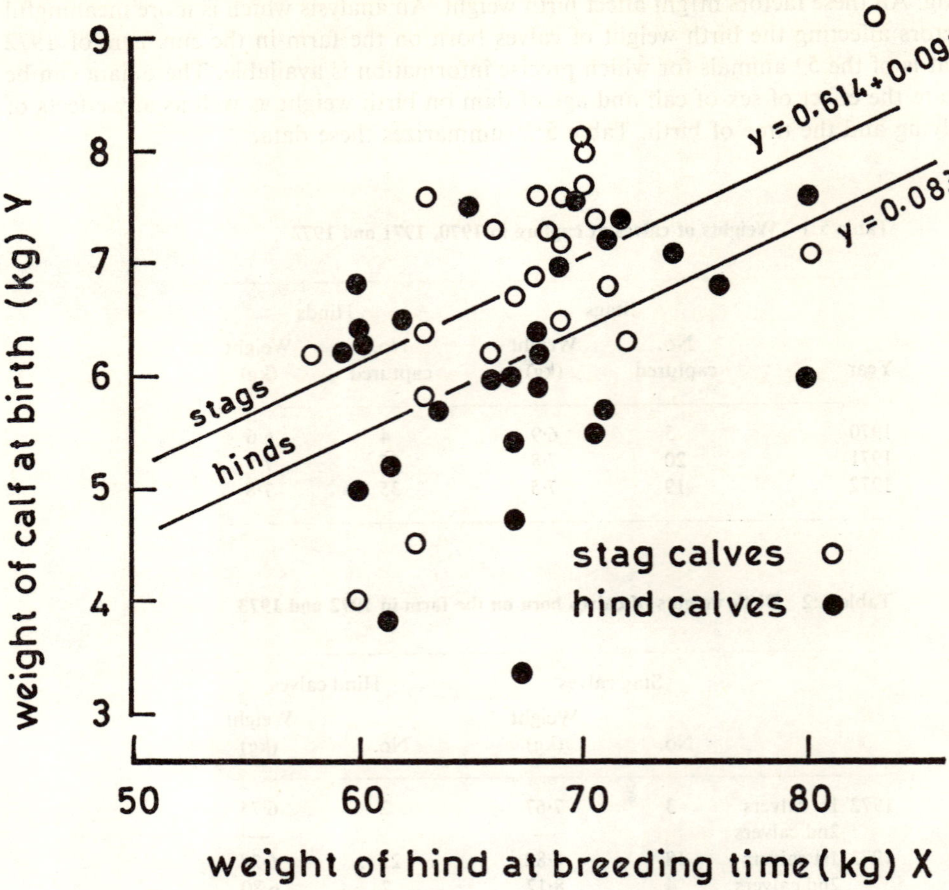

Table 5:3 **Body weights of surviving calves born in June**

Class*	Number surviving	Weight in kg: At commencement	4 weeks	8 weeks	October	December
STAGS						
1970 (R)	4	7·2	11·6	18·8	32·6	43·8
1971 (G)	5	8·6	17·0	29·6	43·6	50·0
1971 (R)	2	8·8	—	19·2	42·0	50·1
1972 (G)	6	8·9	18·6	26·6	41·9	49·8
Home bred 1973	14	[7·0]		—	35·2	41·3
HINDS						
1970 (R)	4	6·6	11·5	15·7	22·3	40·0
1971 (G)	32	7·6	15·0	22·9	31·3	43·5
1971 (R)	17	6·8	—	17·3	27·5	40·3
1972 (G)	23	7·9	16·1	23·8	35·3	41·1
Home bred 1973	18	[6·4]	—	—	32·1	36·3

*In parenthesis is the place Rowett (R) or Glensaugh steading (G) at which the calves were reared. The home bred calves were naturally suckled.

concentrate mixture 'RDM'. In 1971 the group of deer reared at the Rowett Institute were again given the cow-milk substitute and weaned on to dried grass and mixture RDM as before but those reared at Glensaugh steading were given a proprietary ewe-milk substitute, fortified with vegetable oil and designed to be fed at a higher dry-matter concentration, and were weaned on to artificially dried grass and a proprietary fat lamb fattening pellet. It is evident that both stag and hind calves grew at about the same rates at the Rowett Institute in the two years, but that the calves reared at Glensaugh grew very much better. This could be attributed to the superiority for deer of ewe-milk substitute over cow-milk substitute, a matter not surprising in view of the closer similarity of deer milk to ewe than to cow milk. However, at Glensaugh the deer were multiple-suckled from a hanging bucket and thus had milk available at all times while those reared at the Rowett Institute were fed four times daily from a bottle and feeding was not *ad libitum.*

In 1972 a comparison was made at Glensaugh between the two milk substitutes both fed at the same concentration. This comparison was terminated after 5 days since many of the deer calves fed the cow-milk substitute at this high concentration suffered from severe sickness and diarrhoea, and, as shown in Table 5:4, a large number died. Following careful nursing the survivors in this group were given the ewe-milk substitute but even so the effect of the initial period of feeding the cow-milk substitute was discernible the following September in body weights lower by about 1 kg in both hinds and stags. The giving of cow-milk substitute at the same concentration as the ewe-milk substitute was an error; this provided far more lactose than the calf could deal with. No studies have been made to compare the two substitutes at correct concentrations.

Records were kept of the amount of feed consumed and the conversion efficiencies of the calves can thus be ascertained. These conversion efficiencies largely reflect the nutritive value of the individual feeds per unit weight. The proportions of the three ingredients of the diet—milk substitute, concentrate and roughage—changed as the animals increased in age and weight. During the first 8 weeks of life the 1972 calves consumed 1·24 kg milk substitute, 1·01 kg lamb pellets and 0·11 kg hay for every kg weight they gained. This represents a feed conversion of 2·35 kg total dry feed/kg gain. As judged by the 1970 cohort, after 8 weeks, when the calves received no milk substitute and received concentrates and roughage only, feed conversion was 5·1 kg dry feed/kg gain. If the objective is to achieve weights at 3 months of 30 kg, calves reared artificially each require 20–25 kg of a high-quality ewe-milk substitute, 30–40 kg of a good-quality concentrate diet and 20–25 kg of high-quality roughage.

Table 5:4 **Performance of calves fed cow-milk and ewe-milk substitutes each made up to 25% dry matter**

	Cow-milk substitute	Ewe-milk substitute
No. of calves	20	21
No. of deaths	9	2
Weight of survivors in September (kg)		
Hinds	31·7	32·7
Stags	38·0	39·3

Mortality of captured calves

A number of the captured calves died and Table 5:5 summarizes this mortality in general terms while the causes of the deaths of individuals are summarized in Table 5:6.

Some deaths were undoubtedly associated with the many stresses of catching in the wild, of feeding en route (or the lack of it in 1971) and of being introduced into a strange environment. It is unlikely, because of their fleetness of foot, that any calf was older than five days when first caught, but even so, some of the calves accepted the rubber teat quickly, but others did not suck voluntarily for several days. Although these latter calves were carefully force-fed they did not thrive initially although most ultimately became fully integrated into their group and did thrive. One of the calves in 1970 appeared to be terrified and died during feeding, from asphyxia through inhalation of liquid. Two in 1971 also died of inhalation asphyxia. Two in 1971 and two in 1972 drank only small amounts and died of starvation.

In 1971, when calves arrived in ones and twos from many scattered sources, three were dead on arrival or died soon after from the stress effects of travel. With the 1972 collections, which were arranged in batches and in which the calves were trained to artificial feeding, there were no deaths of this kind.

In 1971 and 1972, during the first few weeks after their arrival, when the calves were being fed indoors, 13 out of a total intake of 129 died of infections such as scours, pneumonia, peritonitis, polyarthritis, multiple abscesses and encephalitis. As previously explained, the underlying cause of most of these appeared to be the feeding for a period of 5 days of a cow-milk substitute which was made up with tap water to contain 25% dry matter.

Table 5:5
Mortality of captured stock from time of capture until weaning

	Year		
	1970	1971	1972
Number captured	9	75	54
Casualties			
perinatal causes*	1	9	2
postnatal causes	—	7	10
Total	1	16	12

*Includes deaths during transit or attributable to stress, inhalation pneumonia, starvation due to unwillingness to drink and congenital abnormalities.

Table 5:6 **Specific causes of death in captured calves**

	Perinatal		Postnatal	
Cohort	Calf Ref. No.	Cause	Calf Ref. No.	Cause
1970		asphyxia during feeding		
1971	1F	undersized, stress and starvation	200	chronic pneumonia
	2W	asphyxia	57	rumenitis (engorged on concentrates)
	A	umbilical haemorrhage in utero	19	purulent polyarthritis, pericarditis and encephalitis
	20	dead on arrival—stress	14	choked on straw
	50	brain deformity with atrophy of cerebellum —unwilling to drink	X	not examined
	16	inhalation pneumonia	13	polyarthritis and encephalitis
	72	starvation, slight cerebellum abnormality		
	24	inhalation pneumonia, slight cerebellum abnormality		
1972	SH77	starvation, unwilling to suck	232	scours
	230	septicaemia	SH95	rumenitis and peritonitis
			SH97	scours
			SH96	pneumonia
			SH94	pneumonia
			SH99	no diagnosis
			226	nephritis and abscesses in many organs including brain
			SH98	arthritis and tendosynovitis
			207	bloody guts (haemorrhagic enteric syndrome)

The origin of the final stock

Figure 5:2 shows the location of the estates from which the captured calves which formed the nucleus of the stock were obtained, and these estates are identified in the key to the Figure. Our indebtedness to the owners, factors and staffs of these estates cannot be overestimated. It is at once evident from the Figure that the stock of the farm has been derived from many different parts of Scotland. There is evidence from studies of the biochemical genetics of the deer reported in Chapter 11 that some of these stocks must be regarded as genetically distinct. This is despite the policies of many estate owners continued now over periods of up to 100 years of introducing stags from Warnham Park in attempts to improve their stock. It seems highly probable that breeding of the Glensaugh stock can continue for some time without concern for possible deleterious effects of in-breeding.

Figure 5:2 **The source of the calves used to establish the breeding herd**

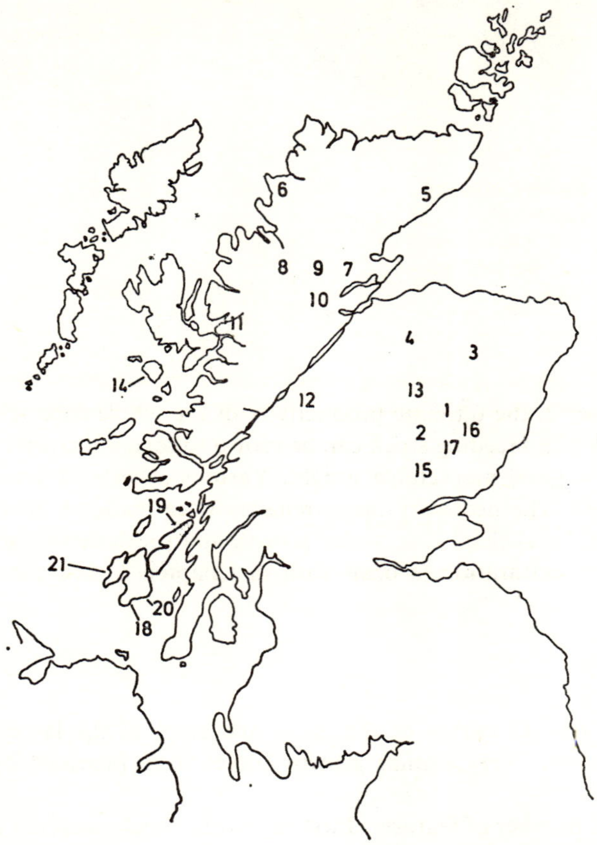

Key

1 Balmoral Estate, Glenmuick, Aberdeenshire
2 Atholl Estates, Blair Atholl, Perthshire
3 Glenavon Estate, Inchrory, Banffshire
4 Dalwhinnie Estate, Inverness-shire
5 Langwell Estate, Berriedale, Caithness
6 Assynt Estate, Glencanisp, Sutherland
7 Wyvis Estate, Evanton, Ross-shire
8 Fannich Estate, Garve, Ross-shire
9 Strathvaich Estate, Garve, Ross-shire
10 Strathconon Estate, Garve, Ross-shire
11 Glencarron Estate, Kyle of Lochalsh, Ross-shire
12 Ardverikie Estate, Laggan, Inverness-shire
13 Gaick Estate, Kingussie, Inverness-shire
14 Nature Conservancy, Rhum, Inverness-shire
15 Forestry Commission, East Conservancy, Dunkeld, Perthshire
16 Dalhousie Estate, Invermark, Angus
17 Cortachy Estate, Glen Clova, Angus
18 Ardtalla Estate, Islay, Argyllshire
19 Tarbert Estate, Jura, Argyllshire
20 Dunlossit Estate, Islay, Argyllshire
21 Islay House Estate, Islay, Argyllshire

Chapter 6. **Growth and development**

As explained in Chapter 2, the ultimate productivity of the whole enterprise depends to a large extent on the rates at which breeding stock can be grown to sexual maturity and surplus fattening stock grown to reach a good marketable weight. Various aspects of growth and development were therefore recorded. The necessary measurements were made at intervals of either one or two months. Body weight, skeletal size, antler growth and tooth development will be described in this Chapter, sexual maturation is dealt with in Chapter 7, and carcase characteristics in Chapter 11.

Body weight

Figure 6:1 shows the growth curves of the hinds and stags of the largest cohort of captured calves, those born in 1971. Very similar growth curves were provided by the calves born in 1970 and 1972.

The Figure reveals a number of features. First, a growth check occurred each winter and these became more severe as the animals matured. The calves received supplementary pelleted concentrates during their first winter, usually at the rate of 1 lb per head daily from September until April. Although their growth rate slackened, especially during January and February of this first winter, it never reached a standstill and by March hind and stag calves had reached weights of about 45 and 55 kg respectively, having gained some 20 to 30 kg since weaning in the previous September. Supplementary feeding was discontinued in April as new herbage began to grow and the calves were then about 20 kg heavier than are wild calves of the same age. In the second and third winters of their lives the animals either received no supplementary food during winter or at most only a little poor hay during storm conditions. Hinds, and more especially stags, tended to lose weight during this time. This winter check and the fact that it was less in the first year of life when supplementary feed was given might suggest that the quantity and quality of winter forage available to the deer was insufficient for growth or even maintenance. However, the winters of 1971 and 1972 were relatively mild and snow cover was seldom a problem, so that grazing remained accessible and little supplementary fodder was required. To explore further the causes of depressed growth in winter, experiments were made under closely controlled conditions at the Rowett Institute. These are reported in Chapter 11. Briefly they show that stags and hinds fed indoors on freely-available diets of good and constant quality also show a growth check during winter. There is thus reason to suppose that the weight-loss shown by deer grazing during the winter is due at least as much to inappetance as to the inadequacies of the forage available.

The sharp fall in weight suffered by the stags during the rut, amounting to as much as 20 kg between September and November in 2- and 3-year-old stags, may similarly be due as much to an essential inappetance as to the excesses and worries of domestic life. Stags penned individually at the Rowett Institute, receiving good food *ad libitum*, show a sudden loss of appetite in early October, even when no hinds are present, and carcase dissection studies show that the weight of food in their fore-stomachs may fall by as much as 10 kg. Yearling stags running with hinds also lost a little weight during the rut, but much less than older animals, while yearling stags isolated from hinds lost no weight at this time.

The eight 1970-born deer made little or no net gain in weight during their third year, from which it may be inferred that they had achieved their mature sizes which were 70 to 90 kg for the hinds and 80 to 100 kg for the stags when at the peak of condition. Wild deer reach similar weights, but they take 4 to 6 years to do so. There is thus no indication at present that the rapid early growth of the Glensaugh deer makes much difference to their ultimate size.

Figure 6:1 **Growth in body weight of hinds and stags captured in 1971**

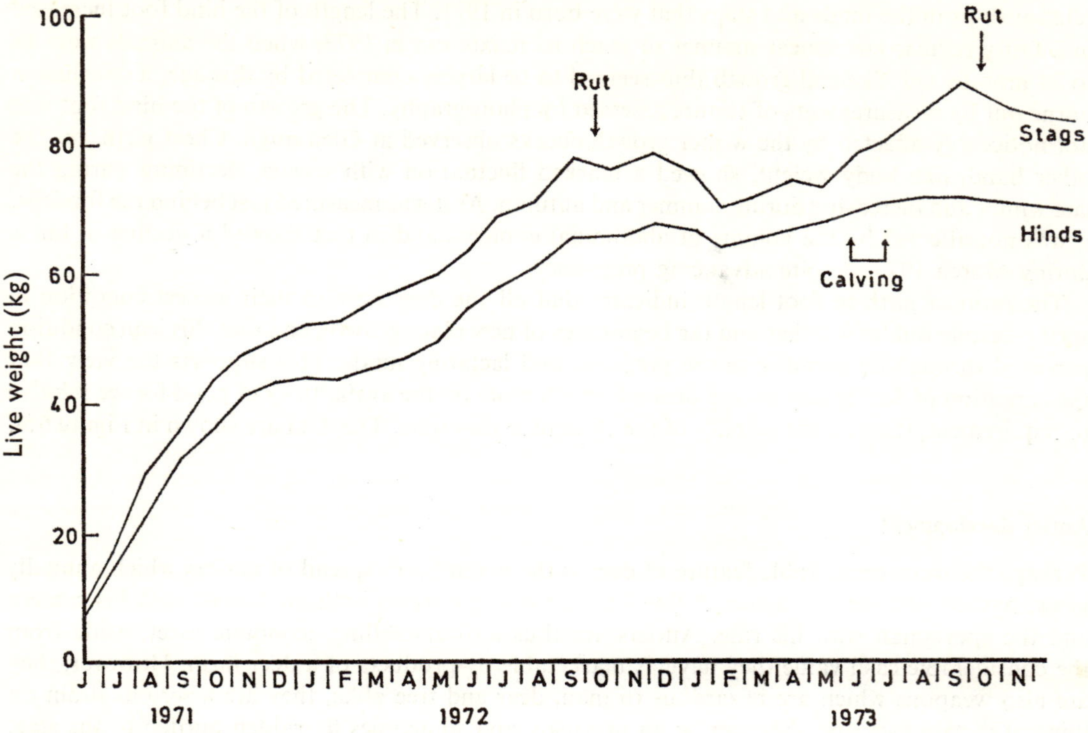

The rate of growth during the summer was uneven. Typically, if the pregnant hinds are excluded, growth accelerated from March to June, slackened during July, especially in stags, and increased again markedly in August and early September. The food intake of stags penned at the Rowett Institute and receiving food of good and constant quality to appetite showed a similar pattern, including the reduction at midsummer which registered on their weight record at the subsequent weighing in July. Stags out of doors are mercilessly pestered by the head fly on warm summer days but this cannot account for the transient reduction of appetite in the housed deer at this time.

The stock as a whole reached its peak of body condition in early September. By then the stags had completed their summer growth and they did not equal this condition again for many months. The hinds at this time were within 1 or 2 kg of the maximum weight which they reach in November or December. This suggests that September might well be the time of choice for slaughter of fat stock. There are other good reasons for this. Removal of surplus stock in September allows calves to be weaned at a viable weight from hinds destined for slaughter; slaughter of stags at this time provides carcases unaffected by 'stag taint'; it allows the remaining breeding stock to sort out their hierarchical relationships and settle down in good time for the rut, and it reduces the number of deer that need to be carried through the winter. An alternative time for slaughter of stags as calves is the end of December, for their early very rapid growth allows weight gains of up to 50 kg in six months. However, the law relating to open and close seasons at present restricts the choice of times for slaughter of deer, domestic or wild, to 1st July–20th October for stags and to 21st October–15th February for hinds.

The deer gained only some 5 to 10 kg in weight between 15 and 27 months of age, a poor return for a year's keep. There is therefore little point in delaying slaughter of surplus stock beyond 15 months of age, unless it is of importance to obtain a crop of calves from the hinds before slaughter.

Skeletal growth

On alternate months from November 1971 onwards the heart girth and the length of the hind foot were measured. It was hoped to establish a clear relationship between girth and bodily condition in the deer, as it is much easier to carry a tape measure to measure girth than a balance to measure body weight. The metatarsus accounts for most of the hind-foot length and its growth is indicative of development of the long bones. The ratio of girth or weight to hind-foot length thus serves as a measure of condition in much the same way that weight-for-height relationships indicate leanness or obesity in humans.

Figures 6:2 and 6:3 show these skeletal measures plotted against age and illustrate their relationships in the hinds and stags that were born in 1971. The length of the hind foot increased in a fairly regular curvilinear manner to reach its maximum in 1973, when the animals were 18 to 24 months old. Skeletal growth thus seemed to be largely completed by this age, a conclusion borne out by measurements of stature assessed by photography. The growth of the hind foot was not noticeably affected by the winter growth checks observed at Glensaugh. Chest girth, on the other hand, like body weight, showed a marked fluctuation with season, declining during the late winter and increasing during summer and autumn. As it was measured just behind the forelegs, it was not affected by the volume of abdominal contents and in fact showed a decline in hinds during March 1973, despite advancing pregnancy.

The ratio of girth to foot length indicates that all the deer were in their leanest condition in spring despite milder weather and the beginnings of new plant growth, and that this lean condition persisted throughout summer in the pregnant and lactating hinds. This supports the view that the condition of deer at any season depends much more on the availability of good forage relative to requirement, than on the severity of the climate at that time. The data are shown in Figure 6:4.

Antler development

Perhaps the most remarkable feature of deer is the magnificent spread of antlers which annually grow, harden and are cast again. A fine head on a stag attracts both the tourist with his camera and the sportsman with his rifle. Antlers are thus an outstanding economic asset, aside from the considerable cash value they have as marketable by-products of a deer farm. However, they are also weapons which are hazardous to man, deer and tree alike, they are a serious drain on mineral metabolism and they can be an unwieldy and sometimes fly-ridden burden to the stag. Antler growth is therefore very important in deer management, even when the prime objective is meat production. Quite apart from such aesthetic and practical considerations is the fact that antlers are the only example of a complex mammalian organ that can wholly regenerate; they provide a very accessible tissue in which skin, cartilage, bone, nerve and blood vessel can readily be studied as they differentiate and grow at a phenomenal rate.

Antlers grow as extensions to the antler pedicles, part of the frontal bones, during May, June and July. Testosterone inhibits antler growth. During the early summer the concentration of this hormone in the blood is low, permitting growth to take place. In August an increasing

Figure 6:2 **Growth of the hind foot in hinds and stags captured in 1971**

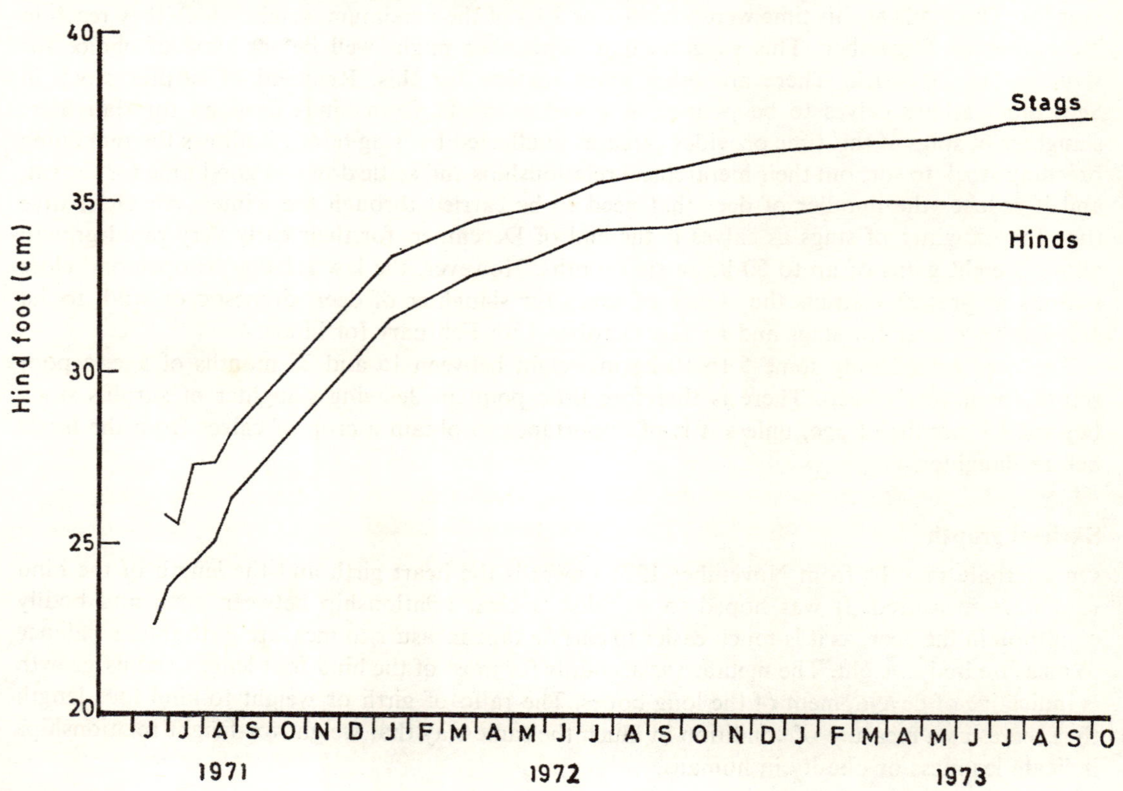

Figure 6:3 **Growth in heart girth of hinds and stags captured in 1971**

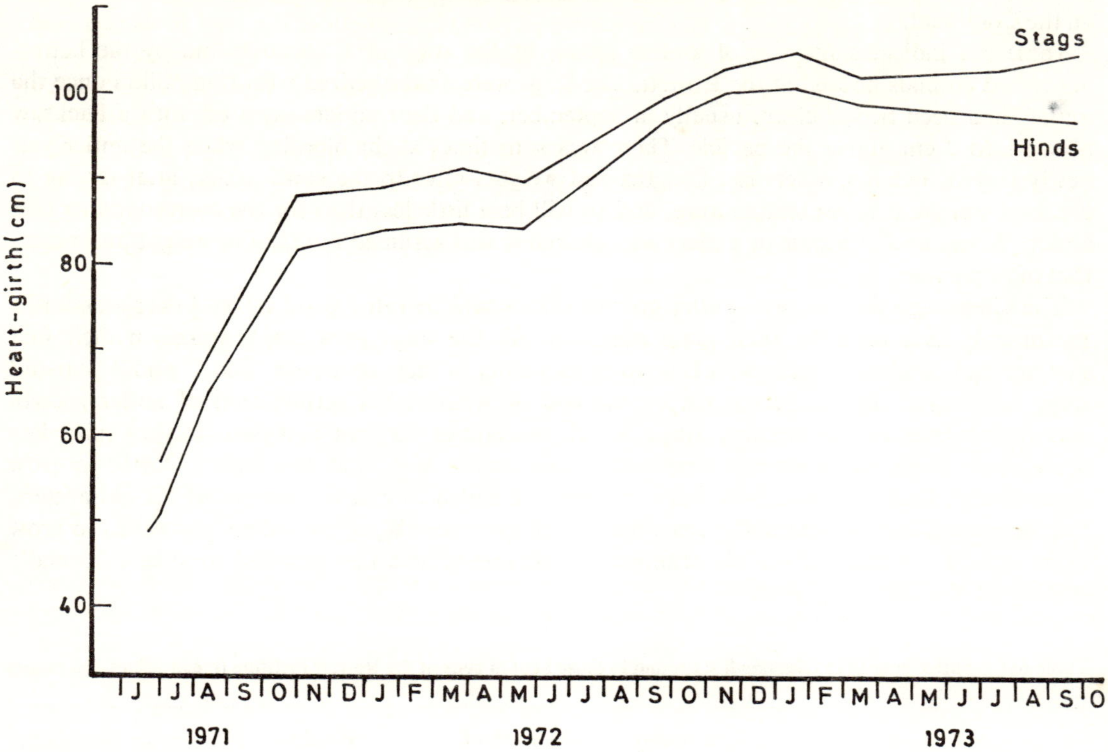

Figure 6:4 **The ratio of girth (a measure of weight) and hind-foot length (a measure of skeletal size) in hinds and stags captured in 1971**

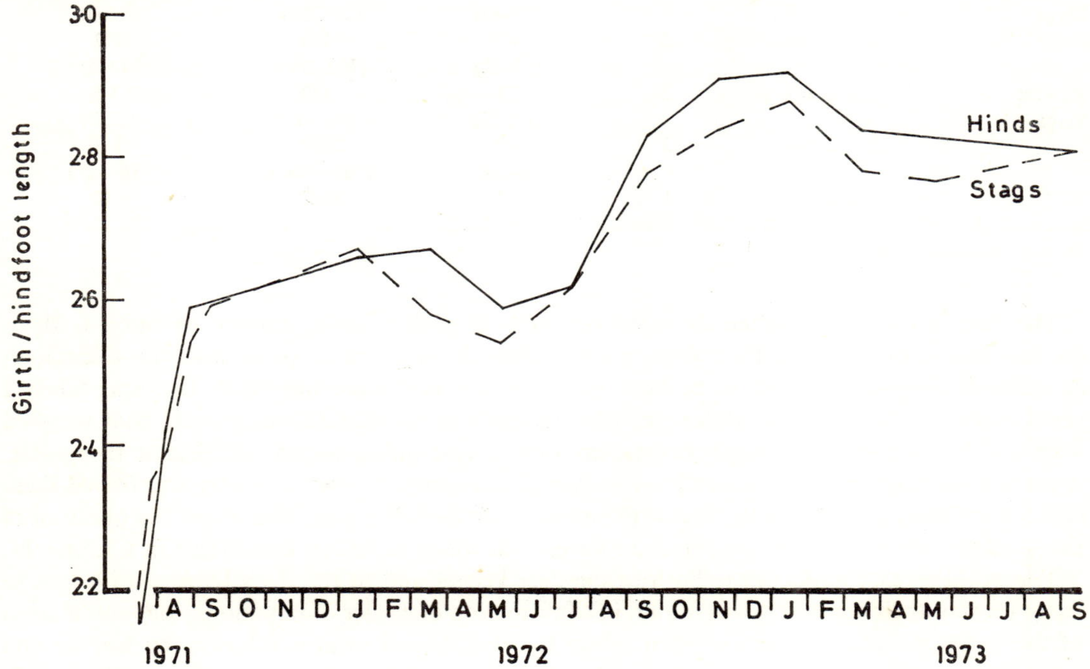

concentration of testosterone causes the skin with which the growing antlers are covered, the 'velvet', to die and it is then frayed off by the stag rubbing his antlers against post or tree to reveal the dead bone beneath. The dead antlers remain in place until falling testosterone levels in April allow antler growth to recommence, starting with resorption of bone at the base of the old antlers which are then cast. The deer normally chew up their cast antlers, thus thriftily recycling their bone minerals. The details and mechanism of antler growth, so far as they are known, are described by Goss (1970).

Antlers increase in size and in the number of tines or points they bear as the stags mature, usually reaching a maximum development at 5 to 8 years of age. Nutrition also influences this process. Stags on Scottish hills generally have much poorer heads than stags in deer parks or on the Continent.

Table 6:1 indicates the size of antlers grown by the stags at Glensaugh and by the better-nourished animals housed at the Rowett. The stags were anaesthetized with Immobilon once the antlers had been frayed clean, usually in September, and their antlers sawn off with a hacksaw about 1 to 2 cm above the pedicle. There was sometimes slight bleeding when the antler was newly frayed, but not otherwise. Lengths and weights refer to the sawn antler, after drying to constant weight at room temperature, and so will be a little less than for the corresponding cast antler. Where one of a pair of antlers was broken it was assumed to equal in weight and points that of its partner.

The Glensaugh deer showed antler growth that would be rated good by wild deer standards, presumably as a result of their good nutrition. All the stags grew single spikes in their first summer and produced heads with four to eight points in their next year. The 2- and 3-year-old stags had frayed their antlers clean by the end of August, but termination of antler growth was rather later in the yearling stags, which sometimes did not complete fraying until late September. The antler remnants were mainly cast during May, and new antlers started to grow immediately. Only one stag, S18, born in 1971, has failed to cast the stumps of his old antlers, but these could not be detached, even when struck quite forcibly. New antlers proceeded to grow from around the sides of the old stumps. The double antlers that resulted were cast normally in May of the following year.

Table 6:1 **Antler growth in Glensaugh stags and in those kept in pens at the Rowett Institute (mean values and range)**

	Age	Number of deer	Length of main shaft (cm)	Weight (both antlers) (g)	Number of points (both antlers)
Glensaugh	1 y. 2 mo.	10	25	110	2·1
Stags			(17–42)	(20–250)	(2–3)
	2 y. 2 mo.	5	47	450	5·2
			(42–52)	(380–580)	(4–8)
Rowett	1 y. 2 mo.	2	23	350	3·5
Stags			(14, 32)	(190, 510)	(2, 5)
	2 y. 2 mo.	3	40	2030	8·7
			(30–50*)	(860–4100)	(6–12*)
	3 y. 3 mo.	1	72	1800	10

*Some antlers misshapen.

The deer at the Rowett often damaged one or both antlers during growth by catching them on the netting or doors of their pens, so it is difficult to compare them with the Glensaugh animals. Their antlers tended to be both more massive and more branched. One stag when 2 years and 2 months old had antlers, misshapen through damage during growth, that weighed 4 kg and had a total of 12 major points; another, a year older, became an elegant 10 pointer. A number of stags housed indoors grew antlers precociously during the winter and frayed them in April or May, only promptly to cast them and grow a second pair. This raises the question of the seasonal, photoperiodic control of antler growth which is further considered in Chapter 11.

The problem that antlers pose both to the deer keeper and to the stags themselves led us to try to prevent their growth. It was found, in confirmation of earlier observations, that amputation of the antler in velvet above the point where the pedicle will form is followed by healing and immediate regrowth. Studies reviewed by Goss (1970) have shown that amputation of the pedicle itself, in stags over a year old in which this structure has differentiated, is attended by little success. However, bearing in mind the differing effects of castration before and after initiation of pedicle formation in the first year—before any antler has appeared castration wholly suppresses antler growth whereas afterwards it leads to the persistence of antlers in velvet—it was decided to test the effects of amputating one of the first pair of antlers, while in velvet and only half grown, within 1 cm of the skull so as to remove the differentiating pedicle. Three animals were chosen and the operation, under Immobilon, is a minor procedure. Rapid healing occurred, but there was no further growth of antler on the amputated side in the year of the operation, and by the summer following there was only some slight bony enlargement under the scalp at the site of the operation. It remains to be seen whether or not antler growth will occur at a later stage.

Hormonal changes associated with antler growth

Analyses of thyroid glands obtained from red deer on the Isle of Rhum have indicated that the concentration of calcitonin in the glands shows two distinct peaks during the year (Phillippo, Lincoln and Lawrence, 1972). These occurred during June and July and during September and October. The rise during the summer was directly correlated with the body weight of the animal and coincided with the growth and mineralization of a new set of antlers. The second peak occurred during the time of the rut.

Plasma samples were taken each month from four stags at Glensaugh over the period from December 1971 to December 1972 for measurement of calcitonin and testosterone levels by radioimmunoassay (Mr. M. Phillippo, personal communication). These were estimated by Dr. L. J. Deftos, San Diego, California and Dr. R. M. Moor, Cambridge. Highest levels of calcitonin, 1400–1900 pg/ml, were found in samples taken during June and July when mineralization of the antlers was occurring. The levels declined to 900–1100 pg/ml during August and early September and showed an increase in late September and October to 1200–1500 pg/ml. These results together with the earlier observations suggest that the thyroidal calcitonin concentration reflects directly the rate of release of calcitonin into the blood.

The plasma testosterone concentrations were quite variable throughout the year even within one animal. A small peak occurring during March and April (200–2400 pg/ml) was followed by a fall to very low levels during May, June and July, and then by a second rise to very high concentrations during August, September and October. Throughout the year, therefore, there did not appear to be any consistent relation between the plasma concentrations of the two hormones.

Development of permanent dentition

At monthly gatherings records were made of the number of permanent incisors which had erupted. These data do not indicate precisely the time of eruption because of the long interval between successive examinations. Nevertheless they provide a useful measure which also illustrates the extent of individual variability. An odd number of permanent teeth was seldom seen, and then for only a short period of time; the teeth thus usually erupt in pairs as in other species. In the analysis of the data it has therefore been assumed that the presence of one erupted incisor indicates that the eruption of its partner is imminent. A calf with three erupted teeth is thus recorded as having two pairs of incisors.

The results of the examination are given in Table 6:2. The mean age for eruption of the first pair of permanent incisors was 12·4 months and for the second 18·4 months. However, the range of ages at which these incisors first appeared was so great, extending to as much as 10 months, that no useful estimate of the age of a deer calf could be derived from the number of erupted incisors. Part of the reason appeared to be an association of eruption with rate of growth of calf. This is shown diagrammatically in Figure 6:5. At a given age there was a spread in the number of incisors erupted, 0 or 2 for 12-month-old calves, 2, 4 or 6 for 17-month-old calves and 4, 6 or 8 for 19-month-old calves. In each age group it was the heavier calves which erupted most incisors. The interval between eruption of successive incisors was fairly constant, at slightly over 6 months. All permanent incisors had erupted by 26 months of age. There was no consistent difference in tooth eruption between stag and hind calves.

Table 6:2 **Dentition of deer at different ages: percentage of animals in age classes with particular numbers of pairs of erupted incisors**

Pairs of permanent teeth	Age in months								
	10	12	13	14	17	19	21	22	26
0	85·7	68·4	39·5	2·6					
1	14·3	31·6	60·5	94·8	38·7	3·6			
2				2·6	61·3	74·6	30·9	37·5	
3						12·7	34·6	12·5	
4						9·1	34·5	50·0	100·0
No. of deer examined	56	38	38	38	93	55	55	16	55

To supplement these observations at Glensaugh, the eruption of incisor teeth was studied by Mr. G. Wenham by taking X-rays of the lower jaw of two stag calves (Plate 10). These animals, housed at the Rowett Institute, were given restricted concentrate rations which allowed them to gain weight about as fast as the Glensaugh calves. The observations are summarized in Table 6:3. It can be seen that mineralization and eruption of the first permanent incisors take some 12 months and that they only rotate to their final position during or shortly after eruption.

References

Goss, R. J. (1970). Problems of antlerogenesis. *Clin. Orthop.* **69**, 227–238.

Phillippo, M., Lincoln, G. A. and Lawrence, C. B. (1972). The relation between thyroidal calcitonin and seasonal and reproductive change in the stag (*Cervus elaphus* L.) *J. Endocr.* **53**, xlviii–xlix.

Figure 6:5 **The relation between weight and age of deer and the number of permanent incisor teeth erupted**

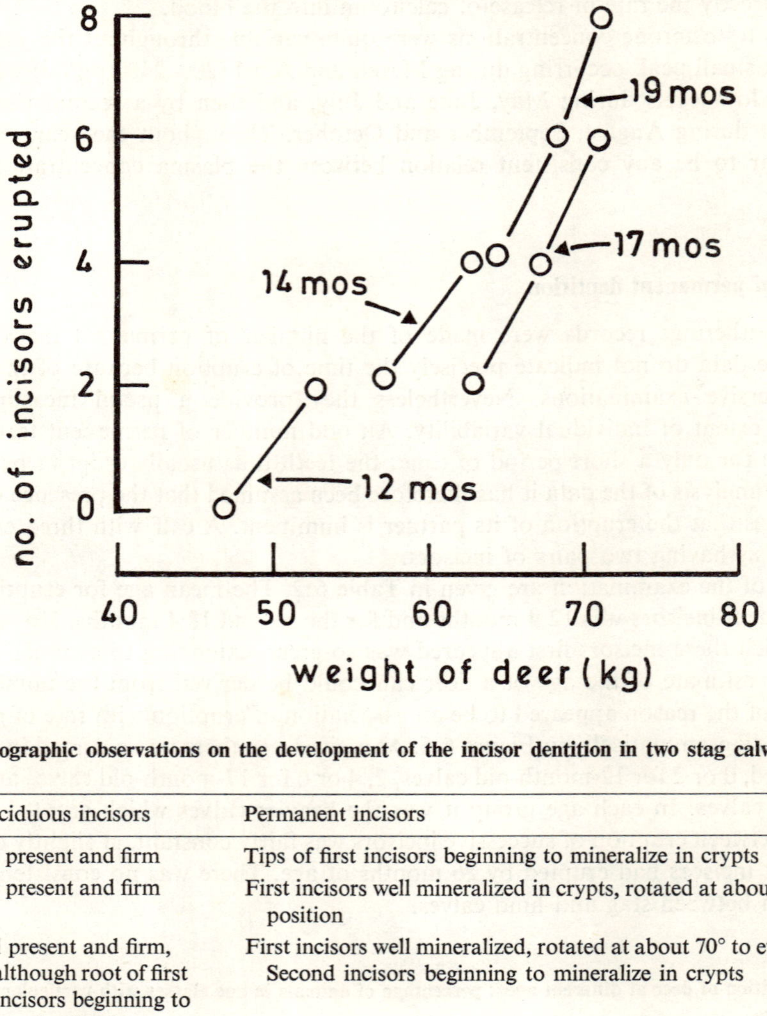

Table 6:3 **Radiographic observations on the development of the incisor dentition in two stag calves (see Plate 10)**

Age	Deciduous incisors	Permanent incisors
3 months	All present and firm	Tips of first incisors beginning to mineralize in crypts
6 months	All present and firm	First incisors well mineralized in crypts, rotated at about 90° to eventual position
9 months	All present and firm, although root of first incisors beginning to resorb slightly	First incisors well mineralized, rotated at about 70° to eventual position. Second incisors beginning to mineralize in crypts
12 months	First incisors shed. Remainder still firm	First incisors fully erupted; fully rotated to final position in one stag but with 30° still to rotate in the second. Second incisors well mineralized in crypts

2. General view of Glensaugh Experimental Farm. (See p. 18)

3. Handling facilities. (See p. 29)

4. Enticing and leading the deer. (See p. 57)

5. Weighing facilities. (See p. 32)

6. Circular crush. (See p. 30)

7. Calf rearing: multiple teat bucket. (See p. 39)

8 (a). Supplementary feeding of calves. (See p. 42)

8 (b). Supplementary feeding of hinds. (See p. 60)

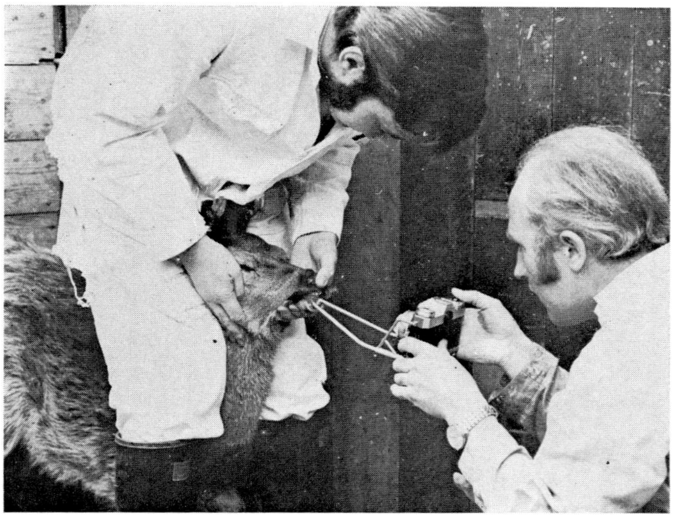

9. Photographing the incisor teeth. (See p. 47)

10. Radiographs of teeth of a stag calf illustrating stages in the mineralisation and eruption of the permanent dentition. (See p. 47)

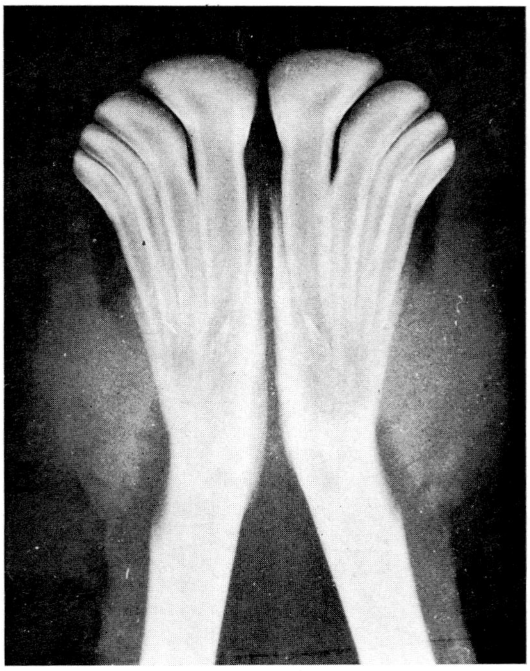

3 months old

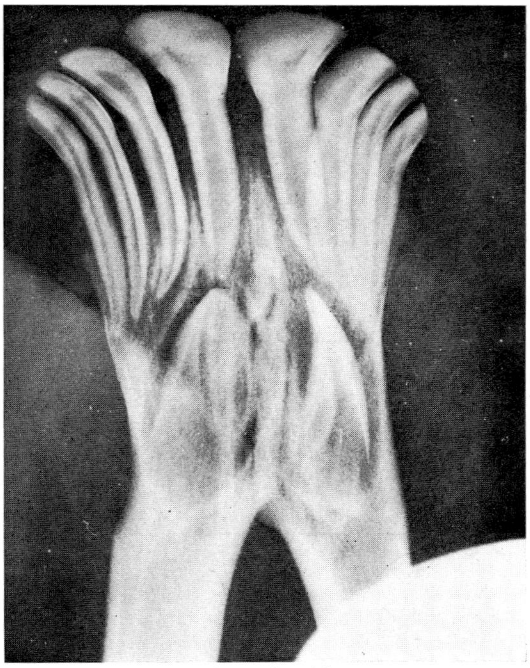

6 months old

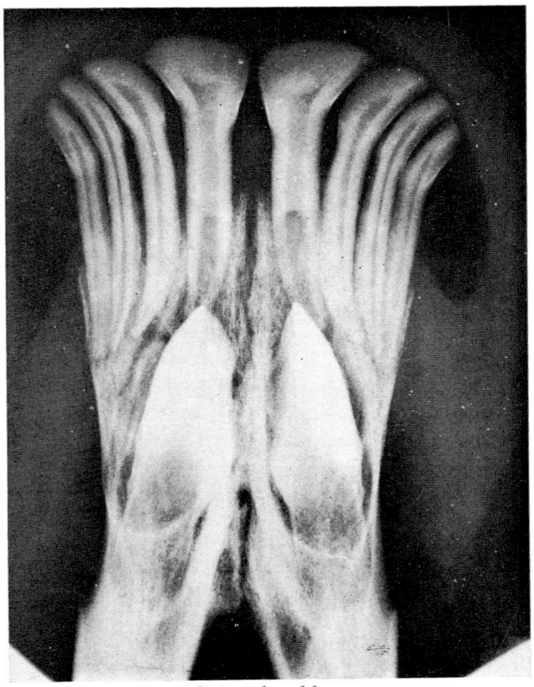

9 months old

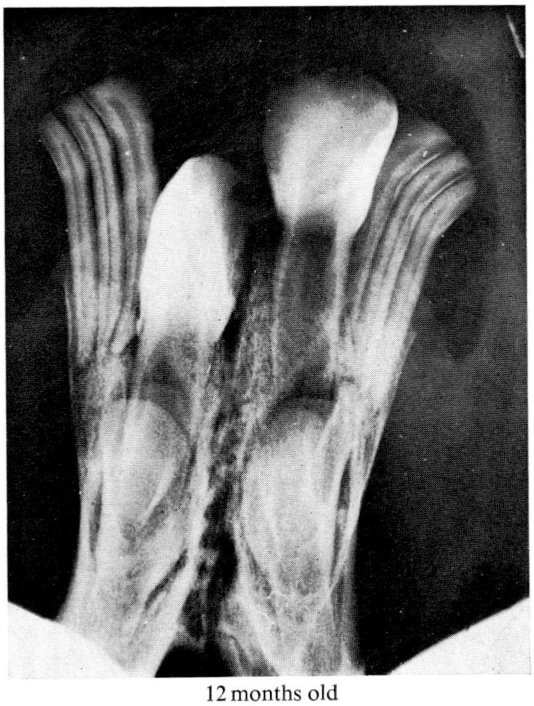

12 months old

11. Calves responding to a call. (See p. 58)

12. Growth of antlers. The antlers are sawn off as soon as the velvet has been stripped, to avoid injury to people. (See p. 46)

13. The head fly, *Hydrotaea irritans*. (See p. 73)

10 cm.

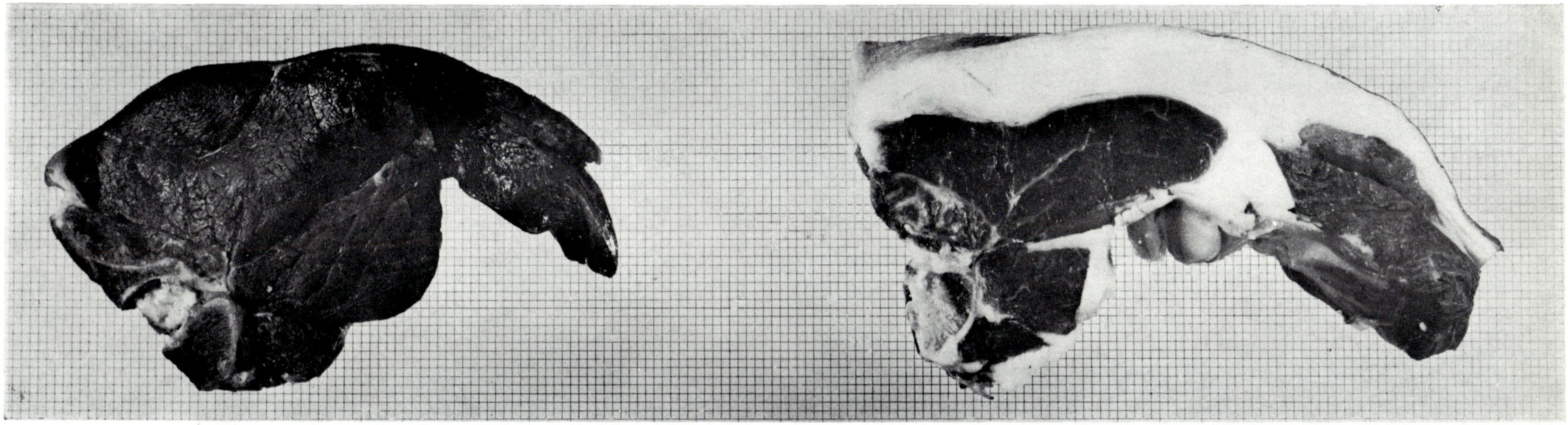

DEER 212 : 11 mo. : 46·5 kg : 14·5·71

LAMB : 10 mo. approx : 45 kg.

14. Loin chops. (See p. 83)

Chapter 7. **Breeding of the farm stock**

The breeding season of 1971

In September 1971 the stock of breeding age on the farm consisted of four hinds and five stags. The hinds ranged in weight from 66·5 kg to 77 kg and the stags from 71 kg to 87 kg. The deer were all placed in a 42-acre paddock for the rut on 14th October 1971, which was rather late in the season, and were left as a group throughout the winter. The group of hinds was supplemented with two further animals which had been served at the Rowett Institute and a further two yearling hinds, which were not pregnant and were in poor condition through having been given a restricted diet, were also brought to the farm at this time. These two non-pregnant hinds were chased by the stags unmercifully but neither of them produced a calf. The pregnant hinds were not molested by the stags to any extent. During the rut one stag, S1, soon established dominance and spent the whole of his time attempting to gather the four hinds and keep the remaining stags at bay. He did, surprisingly, allow the smallest stag, stag AM, to mix with the hinds, this stag being the one at the bottom of the dominance order. A favourite method of protection adopted by stag S1 was to herd the hinds into a corner of the fence around the paddock and then guard them. This same behaviour pattern of using a corner has been noted in subsequent ruts. No fighting between stags was ever observed; the protection simply consisted of chasing away any competing stag which approached the hinds. Hind A2, which weighed 77 kg, came into oestrus early in the morning of the 16th October. During the morning she was chased continuously by the stags but failed to stand. By early afternoon she began to stand for short periods and was unsuccessfully mounted by stag AM. During this time the dominant stag, S1, continued to keep the remaining stags at bay, but apparently allowed AM to attempt to mate. In the late afternoon, however, with stag AM standing close by, stag S1 successfully mated the hind which promptly urinated copiously. It was remarkable that immediately after mating the incessant activity so characteristic of the morning and early afternoon died down completely and no other stag attempted to mate with the hind, an observation confirmed on later occasions. After this mating the dominance order changed and stag S3 took over the role of stag S1. A further hind was seen in oestrus on October 19th when a similar behaviour pattern emerged, but she was not mated before darkness fell. The quiet behaviour of the herd the following morning suggests that she was mated during the night. No further hinds were as closely observed.

Hind A2 calved on 8th June 1972 as did hind A4, which was presumed to have been served four days later. The gestation periods were thus 236 and 232 days. The two remaining hinds calved on the 21st June and on the 4th July respectively. The two hinds brought into the farm when pregnant calved on the 3rd June and on the 12th June. All these hinds calved without difficulty and reared their calves successfully, as described later.

Attempts to mate calves

The growth of the calves which had been captured in June of 1971 had proved so rapid that it seemed possible that the better-grown ones might well exhibit an oestrus when four to five months old. A group of 14 of the heaviest animals, averaging 44 kg on the 10th November and ranging in weight from 39 kg to 48 kg, was placed in a 1-acre paddock together with the smallest stag, AM, which weighed 67 kg. The paddock was close to the observation point but no evidence of oestrus behaviour was noted in the calves. Some evidence was obtained that the stag had attempted to mount one of the calves. The stag was withdrawn on December 13th and none of the calves proved to be pregnant. It may be concluded that despite the rapid growth obtained it is unlikely that hinds can be induced to breed when five months of age.

The breeding season of 1972

When a larger population of hinds of breeding age became available in the autumn of 1972, it was decided to attempt to answer questions related to the optimal size of hind/stag groups by placing such groups in separate paddocks. As mentioned in Chapter 2, there was considerable uncertainty about the number of hinds which a stag could serve successfully and despite the successful use of yearling stags during the rut of 1971, there were some misgivings about their ability to mate a large group of hinds successfully. Table 7:1 summarises the plan of the investigation. There were two groups, A and E, in which two stags, one yearling and one two-year-old, were placed with groups of hinds; group A consisted of older hinds and group E largely of yearling hinds. The remaining three groups were arranged to test whether single two-year-old stags could effectively mate 8, 12 or 16 hinds. In designing the experiment, considerable risk had to be taken. Obviously, if reproductive performance in any of the groups proved to be poor, then the build-up of the farm stock would be placed in jeopardy. Accordingly, the experiment was designed to expose the groups of hinds to the stags for a period of 28 days; hinds come into oestrus at approximately 17 day intervals when cycling regularly so every hind would be expected to come into oestrus at least once during this period. Afterwards fresh yearling stags were added to the groups containing more than 8 hinds for a further period of 14 days. At the end of this period, on November 28th, all the animals with the exception of those in group A were mixed together.

Table 7:1 **The rut of 1972: groupings of hinds and stags‡**

Group	Hinds	Stags	Area allocated for rut: Acres	Supplementary treatment
A (Two stags to 6 hinds)	6: 4 2-yr-old 2 4-yr-old†	S16—yearling 76 kg AM—2-yr-old 83 kg	22 (Stag Park)	S16 withdrawn 1st November and S8 from Group E added 14th November
B (One stag to 16 hinds)	15 yearlings + one older unmanageable hind (Hopetoun)*	S3 —2-yr-old 107 kg	53 (Slochd)	2 yearling stags added: S9 (80 kg) and S15 (75 kg) on 1st November
C (One stag to 12 hinds)	12 yearlings	S1 —2-yr-old 105 kg	18 (West Greenshiels)	2 yearling stags added: S16 (74 kg) and S17 (80 kg)
D (One stag to 8 hinds)	8 yearlings	S5 —2-yr-old 92 kg	22 (Panhandle)	
E (Two stags to 17 hinds)	17: 15 yearlings 2 adults†	S7 —2-yr-old 87 kg S8 —yearling 85 kg	21 (East Greenshiels)	S18 yearling stag added 1st November S8 withdrawn 14th November

†These animals were penned animals transferred to the Farm from the Rowett Institute.
‡Groups B, C and E were all run together (45 hinds and 8 stags) as a group on 10th November, to make a herd of 53 hinds and 9 stags: Group D was added on 14th November. Group A remained separate.
*This animal had been caught when adult and kept in a pen at the Rowett Institute. She never integrated with any group and this may account for her failure to breed. She was eventually stalked and shot.

The behaviour of the stags differed from that noted in the previous year. In group A, where there were two stags, dominance was readily established by the older stag, AM. However, he did not serve the hinds and the yearling stag, S16, in fact was observed to mate two of the hinds in this group. Stag AM died after the rutting season and on post-mortem examination was found to have chronic urethritis, which probably explains his failure to mate. In group E the older stag, S7, likewise soon established dominance and stag S8 was chased away entirely although he was little smaller than S7. In groups B, C and D where there was no internal competition, the stags spent considerable time attempting to fight one another through the fences between adjacent paddocks. Their groups of hinds paid no attention to these contests but took advantage of the opportunity to drift off and graze elsewhere. The fence was damaged in these fights sufficiently to make it possible for a stag to pass through it, had this been attempted, but in fact none did so. The stags continued to use corners of their paddocks to herd their hinds, choosing those corners which were distant from the contiguous paddocks.

When, after 28 days, fresh yearling stags were introduced to the groups they could not establish dominance over the two-year-old stags. They were chased continuously and eventually spent their time walking the fences as far away from the hind group as possible. It was evident that yearling stags, even though fresh, could not compete with two-year-old animals. In group A, which consisted of four two-year-old hinds and two four-year-old hinds with their calves at foot, there was no indication of a separation of the hinds into age classes during the rut. In group E, however, which contained two adult virgin hinds and 15 yearlings, the adults never formed a cohesive group with the yearlings. This caused the dominant stag, S7, some difficulty and he spent considerable time in attempting to bring the two groups together. In the later stages of the 28-day period the yearling stag in this group was noted to be lying near to the two adult hinds. It seems probable—though this was not observed—that the hinds had been mated by this time.

Unlike the previous rut, the stags roared considerably. This was not continuous; during most of the rut a day or two of frequent roaring alternated with a similar period of relative quiet. It was only the two-year-old stags which appeared to roar. Roaring thus appears to be related to the age of the stag.

The assessment of successful matings within each group was made by an examination of the calving dates. Conception date was estimated from calving date on the basis of an assumed 233-day gestation, this being the mean value calculated from the 13 observations of date of mating and date of parturition of hinds at the farm. The range of variation was from -9 to $+7$ days, and the mean value agreed with that found in natural populations by Guinness *et al.* (1971). Only in group D were the results slightly equivocal. Figure 7:1 summarises the pattern of calving in June, July and August 1973. In group A, in which there were two stags, all the six mature hinds can be presumed to have been mated during the 42-day period. In group E, which also contained two stags, only two of the 17 hinds—both yearlings— calved outside the presumptive period, and indeed, since these two calvings were but three days late, it is possible that the two stags in group E were as effective as those in group A. In the groups in which only a single stag was used, the results were less satisfactory. In group D, with eight yearlings and one two-year-old stag, there was a considerable spread in the calving, four hinds calving in early July and one in early August, so that this stag appears to have successfully mated only three hinds. In group C, with 12 yearlings to one stag, two hinds calved in early August. The records for this group are not complete since one hind died under Immobilon anaesthesia. In group B, one hind proved to be barren, one calved in late July, and another in late August. Only 12 of the 16 hinds had been successfully mated by the original stag when on his own. These results suggest that a single two-year-old stag placed with a group of hinds cannot be relied upon to mate with them all. It was of interest that two of the eight yearling hinds in group D were observed to show oestrous behaviour in mid-October but did not calve until 256 days later; this suggests that they were not mated during that oestrus period but during the one that followed.

Figure 7:1 **The distribution of calvings in the mating groups given in Table 7:1. Each point represents a calving.**

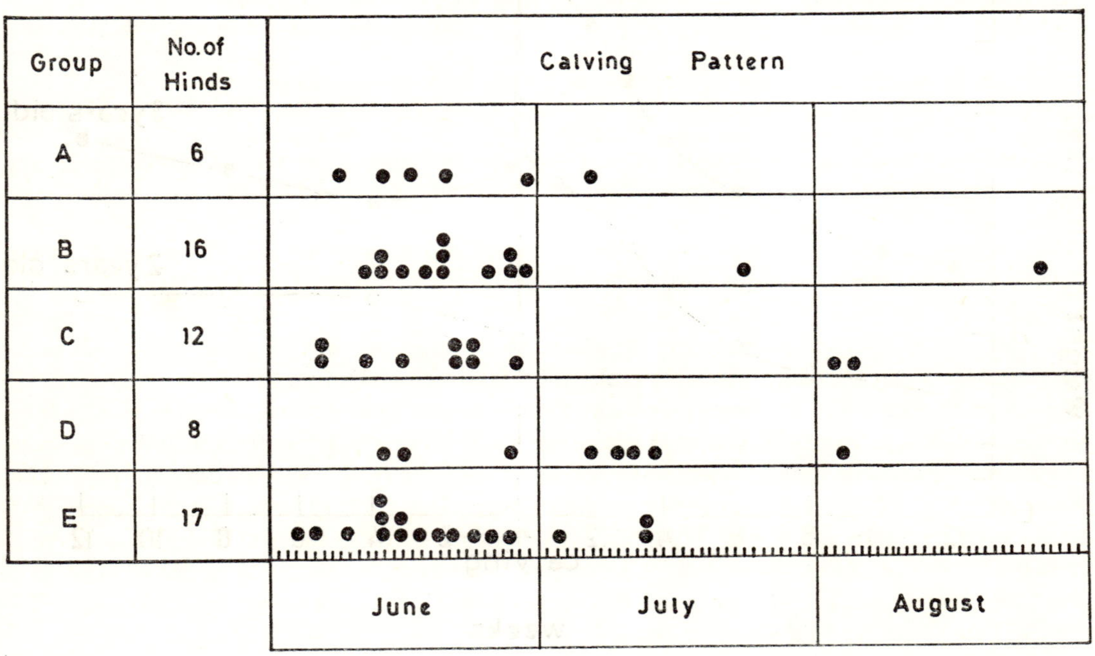

Single tests of this sort are clearly complicated by the variation between stags in libido and fertility. The experiment was repeated in the rut of 1973 and will be reported on in a later volume. Even on the limited information available, however, it would appear that a single stag is capable of serving up to 17 hinds in the season. There is, in addition, the evidence from the behaviour and calving dates of groups B, C and D that some two-year-old stags may need some direct competition by the presence of another stag, even if only a yearling, in the same paddock if they are to settle the hinds successfully during a period of one or two oestrous cycles.

Overall reproductive performance

Table 7:2 summarizes the data obtained on reproduction in the herd for the two years in terms of the number of calves born in relation to the number of hinds put to stag. They show that overall the number of calves born alive to the hinds was 92 per 100 and that the poorer performance in 1973 is largely accounted for by the younger hinds calving for the first time at two years of age. It should be pointed out that in the wild it is very rare for hinds to calve until they are at least three years old. The data already presented in Chapter 5 on birth weight and subsequent growth of calves on the farm born to two-year old hinds show that their growth is satisfactory.

Table 7:2 **Reproductive performance of hinds in 1972 and 1973***

Group	Hinds put to the stag	Number dying before calving	Number barren	Number of calves born			Live births per 100 hinds put to stag
				Alive		Dead	
Calving in summer 1972	6	0	0	6	3 ♀ 3 ♂	0	100
Calving in summer 1973	58	1	1	53	27 ♀ 26 ♂	3	91
Calving at 2 years of age	54	1	1	49		3	91
Calving at 3 years of age	10	0	0	10		0	100
All calvings	64	1	1	59	30 ♀ 29 ♂	3	92

*Hind Hopetoun has been excluded, see footnote to Table 7:1.

Figure 7:2 **Changes in the body weight of two-year-old and three-year-old hinds during pregnancy and lactation**

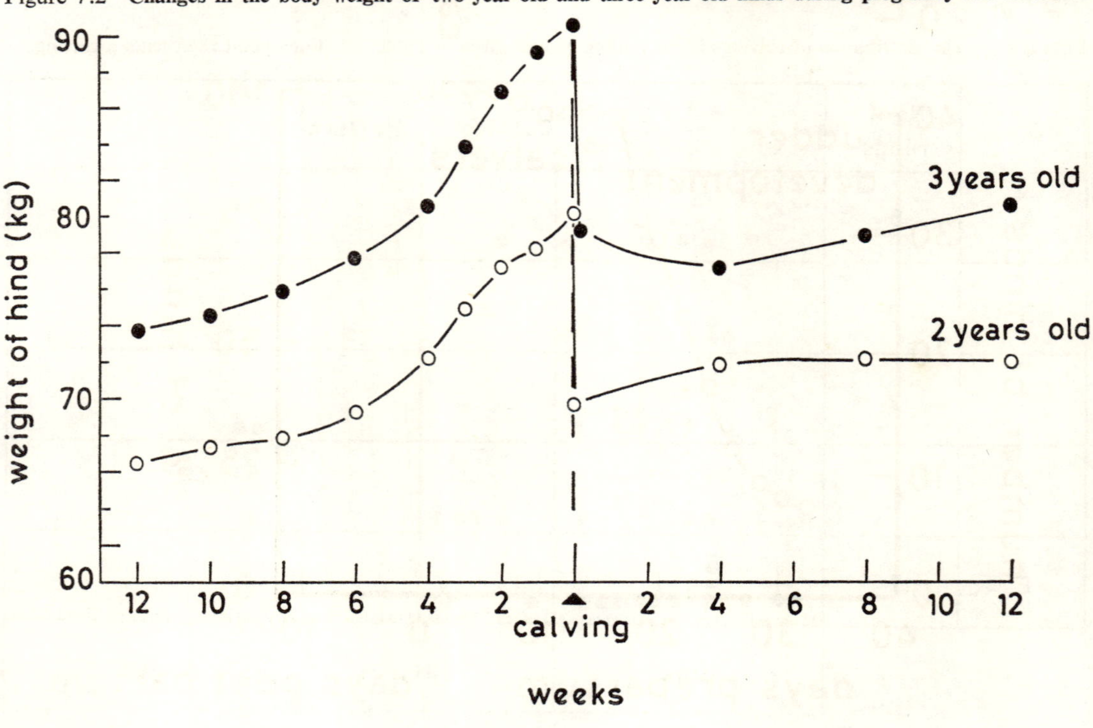

Pregnancy

In the winters of both 1971–72 and 1972–73 no problems were encountered with the pregnant hinds. Supplementary feeding in the form of a pelleted cereal-based concentrate was given to them in both years through May and June at the somewhat arbitrary rate of 1 kg/day.

The hinds were gathered more frequently as calving approached to obtain records of weight changes during pregnancy. These are shown in Figure 7:2. The 37 two-year-old hinds which subsequently lactated increased in weight from 64·4 kg to 80·0 kg—by 15·6 kg, from mid-February to immediately before calving while the three-year-olds increased by 17·2 kg. During the subsequent eight weeks after calving the two-year-olds gained 2·5 kg and the three-year-olds lost 0·3 kg. Whether these changes indicate that pregnancy imposes a greater strain on two-year-old hinds than on older ones and that lactation in the two-year-olds is adjusted downwards to enable the hind herself to grow cannot readily be ascertained from these data. The mean loss of weight on calving in 14 two-year old hinds was 10.4 kg, their calves weighed 6·5 kg which suggests that the placenta and fluids probably weighed 3·9 kg.

Figure 7:3 **The development of the udder in two-year-old and three-year-old hinds before and immediately after calving.**

The scoring system is given in the text. The lower figure gives the number of hinds which showed conspicuous udder development.

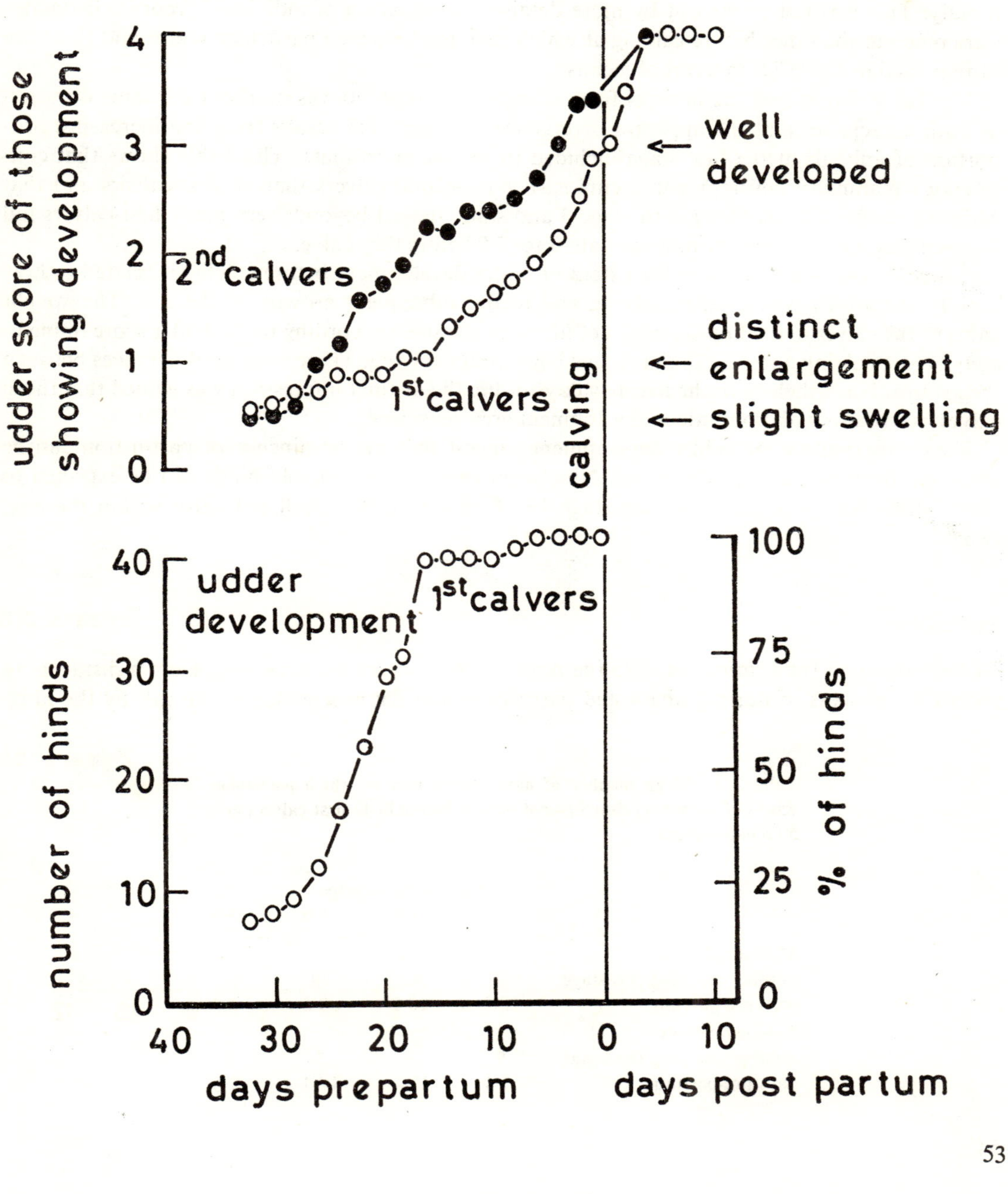

When the hinds were gathered, notes were made on the extent of udder development using a scoring system with the following points.

0 = no udder development

0·5 = four teats beginning to show prominence

1 = distinct udder formation with the udder tissue hypertrophied to a depth of about one inch and the teats prominent

2 = intermediate between 1 and 3

3 = a well developed udder but not of the size seen in full lactation in a first-calving hind

4 = a fully developed udder of large size, as distinct from 3 above

The hinds objected strenuously to palpation of their udders, photography of the udders was not successful and the observations were thus largely based on a purely visual assessment. Gathering was at 3- to 4-day intervals. The data are somewhat complicated by the subjective nature of the assessment of development, and to some extent by the length of the interval between successive observations. Even so, the spread of calving times enabled a reasonably clear picture of udder development to be obtained. Figure 7:3 shows the number of first-calf hinds in which udder development was discernible at different times before calving. At 20 days before calving udder development could be seen in half of the 42 hinds and by 16 days this was evident in all but two of them. In the second-calvers udder development was noted in all five examined on the 31st day before calving. The upper part of Figure 7:3 shows the mean score for those animals in which udder development had been noted, that is for all those animals with a score of 0·5 or more. It will be noted that the first-calvers apparently moved from stage 0·5 to stage 1·0 relatively slowly reaching this at about 19 days before calving, but that second-calvers traversed this interval very rapidly. This was not borne out by more detailed examination of individual records. Estimates were made of the times before calving at which each hind reached particular scores and these are summarised in Table 7:3 in terms of means.

The Table shows that the interval from stage 0·5 to stage 1·0 was of about the same duration in both age classes and the apparent divergence in Figure 7:3 results from the increasing proportion of animals at the first stage included in the mean estimate. The Table shows that conspicuous mammary development occurs earlier in second-calvers than in first-calvers and that while all second-calvers proceed to stage 3 and some indeed beyond, very many first-calvers fail to reach this stage or even the intermediate stage 2·0 before they calve.

It is pertinent to ask whether the extent of udder development before calving is related to how rapidly lactation develops after calving and to the subsequent growth of the calf. The weight gains of the calves in the first six weeks of life were classified according to the udder score immediately before calving and also to that immediately after calving. There were no differences between the groups. Nevertheless, in the first few days of life the distinct impression was gained that those calves from dams with poor udder development grew least well.

These observations on udder development suggest that the imminence of parturition can be predicted from the state of the gland. Ninety per cent of two-year-old hinds can be expected to calve within four weeks of reaching stage 1·0. Conversely, 90% will not calve within the next two weeks.

Parturition

At the calving in the summer of 1973 the hinds were observed continuously, from a distance, to record the process of normal unassisted parturition and the acceptance of the calf by the hind.

Table 7:3 **Mean number of days pre-partum at which particular scores of mammary development were achieved in 42 first-calvers and 5 second-calvers**

	Development stage			
	0·5	1·0	2·0	3·0
42 first-calvers				
number reaching this stage	42	42	39	11
days pre-partum	22·2	18·5	9·0	5·4
5 second-calvers				
number reaching this stage	5	5	5	5
days pre-partum	31·4	27·4	17·4	7·6

The studies were made at the instigation of Dr. Pamela MacDonald in a joint venture by the Nature Conservancy and the two research institutes (Arman, Hamilton & Sharman, in preparation). Watching was accomplished largely by using binoculars by day, and providing light intensification equipment by night. Calving was observed in 26 hinds, and most of these observations were sufficiently detailed to enable recording of the onset of labour and events during labour.

Before calving hinds were noted to become restless and in most instances to move away from the herd. Movement away was naturally restricted by fencing and so many walked the fence while a few exhibited aggressive behaviour towards other hinds at this time. Some bellowing occurred and the hinds were observed to alternate standing with lying, the posture on lying usually being on the side with legs outstretched. Certainly weight was shifted a great deal from stance to stance. The vulva at this time was swollen.

The onset of labour was denoted as the first observation of contraction of the abdominal wall. The intervals between onset of labour and delivery, and between first presentation of some part of the calf and delivery, were about 83 and 36 minutes respectively. However, there was great variation from hind to hind, some delivering their calf as little as 16 minutes after first being seen to go into labour and others taking as long as 173 minutes. In six of the 24 calvings observed the hind delivered her calf from the standing position. The hinds usually licked up amniotic fluid spilt during parturition and often turned to lick the calf's feet while they protruded from the vulva. The placenta was expelled about 91 minutes after birth, and was immediately eaten by the hind.

The calf first stood up about 25 minutes after being born and began to suck its dam's udder some 10 minutes later. In 10 of the 23 instances observed the calf first sucked from the hind while she was still lying down.

At Glensaugh there was one instance of abnormal presentation of the foetus; birth was assisted in this case and in two others in which the hind appeared to have difficulty in delivering the calf. Parturition had previously been closely observed in six hinds confined to pens at the Rowett Institute (Arman, 1974). One of the six gave birth to a dead calf as a result of a posterior presentation, another—again with a posterior presentation—suffered a completely everted uterus, and in a third the process of parturition was prolonged for 4 hours. These difficulties may be attributed to the fat condition and age of some of the hinds as well as to their abnormal surroundings, but despite this bad start the hinds subsequently lactated well (see Chapter 11).

References

ARMAN, PAMELA (1974). A note on parturition and maternal behaviour in captive red deer (*Cervus elaphus* L.). *J. Reprod. Fert.* **37,** 87–90.

GUINNESS, FIONA E., LINCOLN, G. A. and SHORT, R. V. (1971). The reproductive cycle of the female red deer (*Cervus elaphus* L.). *J. Reprod. Fert.* **27,** 427–438.

Chapter 8. **Management and control of the stock**

Without doubt, the success or failure of any deer farming enterprise depends on the ability to manage the stock, that is on the success or failure of a partial domestication of the animals. Unless deer can be moved from grazing to grazing, brought to the handling area for inspection and veterinary treatment and eventually for despatch for slaughter, and separated into groups for any special feeding or manipulation, there is little likelihood that a viable farming system can be entertained. Partial domestication in this sense is distinct from what may be called the production of a tame, petted animal, that is an animal so dependent on man that it has lost or failed to develop those innate characteristics which enable it to survive in a poor environment, even though that environment is necessarily restricted. What is required is that the deer should have a respect for man rather than either an intense fear of him as in the wild or an identification with him as is seen when single infant animals are kept as pets.

The initial bottle feeding of the calves induced a large measure of dependence in them; the first small cohort of 1970 certainly had all the attributes of domestic pets since they received more attention than did the larger intakes of calves in 1971 and 1972. It was not known how far this dependence would survive once the animals were released on the farm, or indeed whether this complete lack of fear and apprehension of man would create problems as the stock aged. The possible behaviour towards man of the second generation was equally unknown; they might revert entirely to a wild state, they might be as amenable as their dams had been as calves, or they might adopt some intermediary behaviour, more compatible with the concept of a partially domesticated stock. The descriptions of behaviour which follow answer some of the many questions which such considerations pose.

Control of the movement of stock

In the first winter of their lives all the calves were hand-fed at troughs in the feeding area on a pelleted diet. Feeding continued at one- or two-day intervals until April. As a result, the calves would come to the sight and call of Mr. Hamilton, who was solely concerned with feeding them. The most effective signal was both to call and whistle, and at the same time rattle a food pail; this would attract deer from a distance of half a mile or more. The effects of this training persisted after trough-feeding had discontinued except in small amounts for the purpose of mustering the deer at the monthly gathering, and when summer herbage had become available for them. Indeed, care had to be taken early in the spring to avoid making too much noise at the handling area otherwise the deer would cease grazing and come in to be fed. As the summer advanced the response to a whistle diminished. The deer would react but took a considerable time before they would move towards the handling areas. During the summer little or no feed was given as a reward for the response to a whistle. As the summer waned there was a distinct tendency for the hinds to become more responsive to a call, but as soon as the mating groups were set up and the stags were placed with the hinds, they refused to respond at all. Clearly, the stag had complete control of his group of hinds. Immediately after mating the hinds once more became responsive to a call. Again no reward was given to the response to a whistle except at the monthly gatherings or when there was a severe storm.

It would therefore seem that with a hand-reared stock which subsequently had been hand-fed for their first winter, handling presented no difficulty. At the monthly gatherings the usual practice was to muster the stock in the morning for it was noted that the animals tended to seek the higher parts of the farm during the late afternoon and become less responsive to a call at this time. It was very difficult indeed to induce the deer to come to the gathering point in the afternoon of summer days.

Stags, when in velvet or after the rut, proved as amenable to a call as did the hinds. During the rut, however, great care had to be taken in moving the tame stags from paddock to paddock for they readily became aggressive to humans. The usual practice adopted was to use a feed bucket containing pellets and to walk down one side of the fence with the stag following on the other. The experience gained in the handling of the stags during the rut suggests very strongly that in the design of the lay-out of paddocks on deer farms one central area should be kept free of stock, not only from the point of view of easing observation, but also from that of moving stags.

Limited experience has been gained so far in the gathering of the second generation of calves born at the farm. The six calves born in 1972 ran with their dams until weaned from them immediately after the rut in mid-December. They were kept separate in a 22-acre paddock from the 1972 intake of hand-reared calves and differences in their behaviour emerged. After weaning they were hand-fed a pelleted supplement and mustered at the monthly gathering in exactly the same way as the calves which had been hand-reared. Even so, the hand-reared stock proved to be considerably less wary of the presence of man, came much more readily to the trough and could, in fact, be handled while standing there. The six naturally-reared calves took some time to learn to come to the trough and always stood off from Mr. Hamilton when he provided feed for them. It was never possible to handle them in the open field. As yearlings, these same animals have also proved more difficult to manage. This has been particularly so of the stags and recourse has had to be made to attempt to drive them into the handling areas for the monthly gatherings. Driving has not altogether been successful; the hinds have usually been gathered without mishap but the stags have, on occasion, broken back and have not been brought in to the handling area. Even as calves running with their dams it was noteworthy that while the first-intake hinds would come to call their calves would hang back.

These limited observations might suggest that the early process of hand-rearing is an essential feature of inducing amenability in the stock. Experience with the much larger number of calves born in 1973, however, does not suggest this to be so. In 1973 the hinds were calved in a smaller area than in 1972 and the calves were weaned and introduced to hand-feeding in September, two or three months earlier than in 1972. In addition, studies of grazing behaviour and of parturition being made during the summer of 1973 ensured some degree of frequent though usually distant contact between the calves and people. This combination of circumstances seems to have produced a group of calves that are about as manageable as the earlier bottle-reared groups. Whether this tameness will persist as well once hand-feeding is discontinued remains to be seen.

Apart from calling-in the animals and driving them with a group of people, no other attempts have been made to use aids to gathering. The accidental introduction of a fox terrier was almost disastrous; the dog was attacked by stags but escaped. Mr. W. Giles of New Zealand, while on a visit to Glensaugh, gave a convincing demonstration of the effectiveness of emulating the roar of the stag. Mr. Giles uses this method to attract hinds in extensive deer-ranching enterprises which he carries out near Wellington. The call attracted deer from all parts of the farm but it may be pointed out that his visit was in early September, immediately before the rut.

There are a number of difficulties with regard to the handling of deer simply by attracting them to a sound associated with feeding. Deer quickly learn to associate the sound with one particular man and the presence of strangers disrupts the behaviour pattern. This has been noted on many occasions when it has proved impossible to bring in animals in the presence of visiting workers from the two institutes. Indeed, it does appear that the deer associate the presence of a white coat or white overall with the gathering procedure itself.

Once the animals are penned they have proved to be very amenable. Unlike similar operations with sheep and cattle, the handling of the animals through a shedder and a race is accompanied by no noise whatsoever by the animals themselves. At these gatherings it is, however, possible to note that some individuals are more difficult to handle than others. For example, stag S7 and stag B7 both proved very intractable both as calves and as adults. Experience has also shown that it is inadvisable to pen animals closely for too long a period. This is particularly true of stags of different ages for the younger and weaker animals have no chance of avoiding the attentions of the more dominant stags. The most dangerous times, of course, are during the late summer, when the antlers are fraying, and later during the rut even if the antlers have been removed. When closely confined even hinds may fight by biting or kicking and can damage one another. The size of an initial holding pen is important for both stags and hinds. Hinds will usually settle without disturbance provided they are given about 60 square feet per beast. The same is true of stags within a single age group. Closer confinement can result in one dominant stag inducing his companions to leap a seven-foot fence.

The problems of mixing stocks of deer

When describing the behaviour of the rut of 1972 mention was made of the problems of integration of older hinds with yearlings. No problems were encountered on mixing weaned three-month-old calves, one group having been reared at the Rowett Institute, the other at the steading at Glensaugh. A number of problems have, however, arisen when animals which had been reared at the Rowett Institute for longer periods of time have been taken to the deer farm. In very few instances has the introduced group integrated with the main stock. It would appear that the mixing of stocks is quite feasible up to an age of six months; thereafter difficulties may ensue. Animals which have been captured in the wild and kept under pen conditions at the Rowett Institute for long periods during which time they have apparently become completely tame, have promptly reverted to their wild state when released on the farm at Glensaugh. For example, one animal referred to in Tables 7:1 and 7:2 as 'Hopetoun' completely failed to integrate with the hind stock, was impossible to gather, became a lone individual running the northernmost perimeter of the farm and eventually had to be stalked and shot. This animal was a park animal which had been subject to experiments at the Rowett Institute, was judged sufficiently tame to be a useful farm animal, and yet proved quite unmanageable. She also failed to conceive during the rut. A second example is that of two hinds which had been kept at the Institute for a period of three years. They had proved to be completely tame and could be led within the Institute grounds for weighing and similar procedures. These animals have been introduced to the farm and although it has proved possible to gather them, they have not been accepted by the hind stock to which they were added and have remained isolated and inconvenient animals.

Restraint

Except for the stags during the rut most simple handling procedures of the deer, such as the reading of ear tags, weighing, faecal sampling, examining for the presence of warble larvae and the taking of certain body measurements, could be carried out in the race with minimal restraint. For blood sampling the crush, briefly described in Chapter 4 and similar to a cattle tuberculin-testing crush, was sometimes used. Though not entirely satisfactory the crush holds the head up and prevents the deer from pulling back.

For more elaborate procedures such as saliva sampling, the paring of hooves, the measuring of testicles in stags, or the machine-milking of hinds, the narcotic-tranquillizer drug combination etorphine hydrochloride-acepromazine ('Immobilon Large Animal') was used. When the procedure has been completed an intravenous injection of diprenorphine ('Revivon Large Animal') antagonizes the narcotic and permits the deer to rise to its feet in 90 seconds or so, though it is still under the effect of the tranquillizer. The use of these drugs, which are manufactured by Reckitt & Colman Ltd., is controlled by the provisions of the Dangerous Drugs Act and, except in very special circumstances, their sale and use are restricted by law to qualified veterinary surgeons.

Immobilon is given by intra-muscular injection with a syringe into the *Gluteus* muscle, taking care to avoid the hip joint. This site was found to be readily accessible and less sensitive than the thigh. The deer do not rear up when injected in this site. The full recommended dose of 1 ml/100 kg body weight has been given except to pregnant hinds which have been given 0·7 ml/100 kg. Three to four minutes after the injection when the deer starts to take short high unsteady steps, the animal has been guided into a suitable place and gently pushed into a sitting position. A minute or so later it is propped on its side with its head suitably cradled to prevent possible inhalation of regurgitated rumen contents. For milking, an intravenous injection of oxytocin is given when the deer is under the influence of Immobilon and milk is withdrawn starting one minute later, either with the forefinger and thumb for small samples or, for complete milking, with a modified sow milking machine (Arman *et al.*, 1974). It has been discovered that although Immobilon has been used repeatedly (as many as 17 times in one hind at approximately weekly intervals), the effective dose relative to liveweight remained the same (Sharman, 1973).

Use of Immobilon has made possible observations which would otherwise have been impossible or too hazardous. It is a great asset to be able to revive the animal in one or two minutes and to turn it out of the handling area to make way for others. A disadvantage, though not a serious one on the farm, is that the animal's appetite is considerably reduced for a few days after the injection. This effect has been examined in controlled tests at the Rowett Institute. Four stags were studied each on 13 to 15 occasions. Comparisons were made of the mean feed intake three days before restraint and three days afterwards. The mean reduction of intake was 36·4%, there being no differences between the declines induced in the different stags. The responses were variable from occasion to occasion, the standard deviation being $\pm21{\cdot}7\%$. Clearly the use of Immobilon and Revivon to study deer has to be with circumspection, for it affects their nutrition considerably.

Surgical castration was carried out on three calves weighing 14 kg each using the Small Animal Immobilon and Revivon preparations of Reckitt & Colman for use in dogs and cats. 2 ml was given by intra-muscular injection instead of the 1·5 ml recommended for dogs of the same weight. After 6 minutes, narcosis was sufficient for the operation to be carried out successfully with only gentle restraint by an assistant. The calves regained their feet after intravenous administration of the antagonist, but remained tranquillized.

Aggression towards man

Considerable care has been taken to avoid any injury to people working at the farm, but even so, a number of incidents have taken place which suggests that liberties should never be taken either with hinds or with stags. Some examples might suffice. Firstly, the tame hind stock reacts aggressively towards strangers. Their behaviour pattern consists of walking aggressively towards the stranger and, on occasion, rising on the hind legs and striking out with the fore feet. This has been noted to be particularly marked if children accompany the stranger. Hinds appear to be particularly attracted to children and it is the children rather than the adults who are attacked. Secondly, a hind at calving can be extremely dangerous. One hind attacked Mr. Hamilton when he was weighing a newly-born calf; the attack was not relinquished despite the use of a stick and it was only because of the close proximity of a fence that escape was possible.

When the velvet of the antlers is fraying and before the rut stags are extremely unpredictable. When a member of staff was making her way through the corner of a paddock containing hinds and a two-year-old stag, the stag which had been very friendly until then, lowered its head and knocked her over. Fortunately it is routine practice to saw the antlers off as soon as the velvet strips off in August-September and the stag had only two short rounded stumps remaining. He boxed her several times with these before someone came to her aid by driving the stag off with a stout stick. The lady had been armed with a stick but she broke it in trying to defend herself.

On another occasion a stag, hitherto quite quiet, charged for no apparent reason. Happily a fence was near and the assistant escaped. A further incident which might have had serious consequences by good fortune did not. It highlights a dilemma which will face all potential deer farmers. One fine autumn morning a lady and six young children climbed through the perimeter fence and walked through a paddock containing hinds and a rutting stag. These people either did not know that there were deer present or more likely were attracted by the roaring of the stags. They were obviously unaware of the danger. The dilemma is whether or not to surround the perimeter of the deer farm with warning notices. Notices can sometimes invite the people they are designed to exclude.

References

Arman, Pamela, Kay, R. N. B., Goodall, E. D. & Sharman, G. A. M. (1974). The composition and yield of milk from captive red deer (*Cervus elaphus* L.). *J. Reprod. Fert.* **37,** 67–84.

Sharman, G. A. M. (1973). Immobilon in deer. *Vet. Rec.* **93,** 114–115.

Chapter 9. **Pasture management and grazing behaviour**

Management of the grazings

It soon became evident that there were a number of restraints upon the optimal use of the grazings on the farm. Firstly, because of the non-central location of the handling facilities, at the monthly gatherings stock had to be held for a while in those paddocks and grazings near to the handling facilities. Secondly, at the time of the rut, all grazings had to be used to contain breeding groups. Thirdly, at calving time further sub-division of stock had to take place to accommodate a spread of calving. Lastly, in the early stages of the project, stocks were far too small to warrant any use of the remoter grazings. The location of the handling facilities and of the divisions also meant that it was not always possible to move stock to the remoter areas in any simple sequence since it is, of course, impossible to move one stock through another. In addition, when deer are hungry, as they are immediately following a monthly gathering, movement through one pasture to another is virtually impossible. The deer cannot be called until they have eaten their fill.

These restraints have meant that it has proved difficult to make what might be regarded as the most efficient use of the various grazings. In general, the remoter grazings have been less heavily stocked than have the nearer ones. The nearer grazings are, however, both the more productive in terms of the keep that they provide during the summer time, and the more sheltered ones to be used in the winter. While these latter considerations suggest that the remoter grazings on the plateau, namely, Birnie and Goyle, could not be expected to provide keep throughout the winter of the year, the restraints mentioned earlier have, in some respects, impeded their use in summer and autumn.

An analysis of the use of the grazings is only informative during the year from December 1972 to December 1973 when the hind stock had reached reasonable numbers. Records were kept of the movements of all stock and from these an estimate was made of the number of grazing days of a hind or her equivalent provided by each grazing. The weighting used to equate for age classes is given in Table 9:1. From the stock records and Table 9:1 estimates were made of the number of days' grazing provided by each paddock every month. These are given in Table 9:2 where they are summated, divided by the acreages of the grazings to give the number of grazing units per acre per year. This number of hind grazing days per acre can be expressed as a stocking rate, that is, the number of acres theoretically required to maintain one deer for a year. These derived calculations are given at the foot of Table 9:2.

It is convenient to group the grazings and consider together firstly Birnie and Goyle on the plateau, secondly the more accessible grazings of Slochd, Panhandle, West and East Greenshiels, and lastly the smaller paddocks comprising the Stag Park, Foxhole and Woodside. Birnie and Goyle provided keep for the deer until early in January of 1972 and they were not used again until March and then but moderately. They received their heaviest grazing during August and September. Altogether these grazings provided less than 40 days of grazing per hind per acre, the corresponding stocking density being one hind to 10 acres. The pressure on these two grazings was thus very small indeed and 1973 was the first year in which they had been subjected to any pressure at all. Understandably, there was no immediately discernible change in the nature of the grazings in terms of their botanical composition. The two grazings could well have carried considerably more stock; by September there was still a considerable amount of keep from ling and the grazings went into the winter in a very good state.

Table 9:1 **Weights used to calculate stocking rates**

Type of stock	One day of grazing = units
All animals aged more than 11 months	1
Calves sucking their dams in June and July	0
Calves in August	0·2
September	0·3
October	0·5
Calves in November, December and January	0·6
Calves in February	0·7
March	0·8
April	0·9

Table 9:2 **The utilization of the farm grazings. The unit in one day's grazing by a hind or 'hind-equivalent'**

		Remote grazings		Accessible grazings				Near-by paddocks				
Month	Days	Birnie	Goyle	Slochd	Pan-handle	West Greens.	East Greens.	Stag Park	Forest	Fox-hole	Wood-side	Total
January	31	627	430	316	238	90	608	131	256	70	0	2766
February	28	0	782	303	290	230	461	267	197	94	0	2624
March	31	606	899	223	232	0	587	335	110	30	0	3022
April	30	911	308	200	334	0	800	285	84	82	0	3004
May	31	163	332	401	399	450	788	311	137	272	0	3253
June	30	800	223	314	311	577	486	270	82	80	0	3143
July	31	340	794	419	256	507	406	279	80	57	114	3252
August	31	592	1078	418	239	260	295	279	189	68	96	3514
September	30	962	1228	216	229	0	30	286	372	212	66	3601
											(264)*	
October	31	435	160	598	294	520	353	590	92	228	336	3870
November	30	1381	0	420	460	310	633	200	106	23	336	3869
December	31	960	560	1044	290	0	762	99	218	46	16	3995
Total per paddock	365	7777	6794	4872	3572	2944	6209	3332	1923	1262	1228	39913
Acreage		200	183	53	22	18	21	22	16	6	3	
Total per acre		38·88	37·32	91·92	162·36	163·55	275·95	151·45	120·18	210·33	409·33	73·5
Units per day per acre		0·106	0·102	0·251	0·444	0·448	0·756	0·414	0·329	0·576	2·224	0·201
Stocking rate per acre		1:10	1:10	1:4	1:2·3	1:2·2	1:1·4	1:2·4	1:3·1	1:1·8	1:0·45	1:5†

*264 grazing units contributed by the feeding area during October while calves were fully supplemented—the area was bare of any vegetation—figure is included to keep the totals correct.

†Stocking rate overall at 31.12.73.

Supplementary feeding:

(1) 5 cwts of hay fed in Stag Park during months of January, February and March.

(2) One ton of barley cubes fed to hinds in May and one ton in June to animals in Slochd, Panhandle, East and West Greenshiels.

(3) Weaned calves were fed 1 lb/head/day during January, February and March, October, November and December in Panhandle, East Greenshiels and the feeding area.

On the accessible grazings of Slochd, Panhandle, West Greenshiels and East Greenshiels, grazing was possible during January and February, indeed, as the sheltered grazings, they provided most of the keep of the stock during this period. They were also heavily stocked throughout the spring and summer, rested to some extent in September and used in October as areas for breeding. Immediately after the rut the stocking of these areas was reduced and Birnie and Goyle were used again. Considering that these accessible grazing units are unimproved hill grazings, their carrying capacity for deer stock was impressive. East Greenshiels was stocked at the rate of one hind or its equivalent to every 1·4 acres, and West Greenshiels at the rate of one hind to 2·2 acres. The Slochd was less heavily stocked at one deer to 4 acres. There was little evidence of change in the characteristics of the herbage on the Slochd grazing, indicating that here stocking intensity was still well below carrying capacity. On the West Greenshiels and East Greenshiels, however, there was evidence of botanical change. These two grazings had also been stocked heavily during the two preceding years when it was thought desirable to keep the stock under observation from the handling area as much as possible. The changes in the composition of these grazings consisted in the main of a marked reduction in the amount of ling and the development of a grassier sward consisting of *Deschampsia flexuosa*, *Agrostis* sp. and *Festuca* sp. together with *Juncus squarrosus*, *Luzula campestris* and some sedges. In addition, there was a marked reduction in the bracken coverage and a thinning of the bracken stands with the emergence of small areas within the bracken colonized by grasses. Deer will eat a little bracken in the early growing stages and also when the fronds are fully emerged. Then deer were seen to graze to a limited extent on the frond itself. They also used bracken for cover, particularly when the fly infestations, common in June, July and August, caused them irritation. Here they may well also graze some bracken but evidently seek grasses which are growing in the stand. No cases of bracken poisoning have been diagnosed but some cases of impaction, which resolved themselves without treatment, have been noted in calves when first turned out to the Foxhole when it was noted that a few animals ate bracken fairly avidly.

The extent of grazing in the near-by paddocks was largely determined by management practices and the need to keep young calves under surveillance.

Grazing in the Forest area

It will be noted from Table 9:2 that the Forest area was relatively heavily stocked at one deer to 3·1 acres. The Forest area is divided into two, the ride to the Stag Park and the plantation proper. During the period from October 1972 to February 1973, the numbers of days on which individuals had access to the stag ride were 134 'stag days', 632 'hind days' and 455 'calf days'. In the same period the plantation paddock provided 242 stag days and 423 hind days. The greatest use of the plantation was made in January and February.

A severe gale in February 1972 blew down about 200 trees. Most of these—85%—were lodgepole pine and the remainder Norway spruce. Inspection of the trees and observation of the deer showed that the most of the accessible leading shoots of the blown trees had been browsed, but browsing of the pine was less than that of the spruce. At this time a little damage to the standing trees was noticeable. About 3% of trees had been damaged, this consisting of signs of nibbling and rubbing of bark thus exposing wood. Other trees had been rubbed sufficiently to cause resin to flow. No consistent distribution pattern of damage nor timing of damage has so far been noted, neither have any individuals, prone to causing damage, been identified. Observations are continuing.

Grazing behaviour

The behaviour of animals within any paddock is of interest. The general observations of grazing behaviour made during the early years of the work were reinforced in the summer of 1973 by a detailed study of the grazing behaviour and movement of 16 hinds in East and West Greenshiels. From this study and from general observation it is clear that deer rapidly seek particular areas of flush herbage and graze them heavily while largely avoiding other areas where there is limited herbage. Grazing is inhibited by high winds when deer seek the shelter provided by the lie of the land and this is particularly apparent when the wind is cold or drives rain. The grazing pattern of the deer was unlike that of hill sheep and to some extent unlike that of downland sheep in that they grazed as a herd rather than as a flock. That is, they tend to graze fairly closely together, often in line abreast, separated by a few yards with leaders in front and the rest behind. If there are two social groups these graze independently. However, the deer were similar to hill sheep in ranging the hill from a night rest area to the valley in the morning with the reverse movement at night. The detailed study, which involved plotting hind movement and activities in the 22 acres of Greenshiels, showed that few areas were not traversed by the deer and that the distance moved per day by a hind in full lactation was at least 1·5 km in the hours between dawn and dusk.

In all the grazings it was found that the fences are walked by the stock. This is general throughout the year but comes to a peak during rutting when the stags, particularly those low in the dominance order, walk the fences fairly continuously; it also takes place during calving time when hinds, close to calving, also walk the fence.

The behaviour of the deer in relation to both favourable and unfavourable weather conditions is of interest. Firstly, the deer were observed to have a 'sunning' behaviour pattern. They appeared to be highly appreciative of sunshine and during sunny periods they would lie flat on their sides, legs outstretched with the belly exposed to the sun. There is some evidence too that they tend to seek south-facing aspects which catch the sun, this being more apparent in windy weather. The farm has not experienced very much snow during its first three winters. Deep snow presents few problems for red deer in the wild and this was equally so of those which had been orientated to the farm steading. These long-legged animals came through quite deep snow with ease and were then hand-fed. During the winter of 1972–73 four hinds which had been introduced to the farm, having been penned previously at the Rowett Institute, did not come in during a heavy snow-storm but remained at 1250 ft at the top of the Panhandle. Two of them made circular holes in the snow some 5 ft in diameter, the snow at this point being some 4–5 ft in depth. They lay in these holes and refused to come out. They were fed there for two days when an attempt was made to dig them out. They refused to come to the steading and returned to their pits and were again fed there.

Feed selection

In the summer of 1973 a study was also made of the eating behaviour of two hinds. It is extremely difficult to obtain estimates of what is eaten by a grazing animal even when it grazes an even stand of a pasture consisting of but one or two species of grass and clover. When the sward is highly complex and growth is variable the problem is even more difficult. A simple but extremely demanding approach was used; from dawn to dusk every single bite taken by a single deer was recorded and the plant species taken at each bite was identified. The investigation, ably carried out by Miss Ineson and Miss Mohun, entailed recording nearly one hundred thousand bites. In more detail, one or other of two unusually tame hinds was followed closely from dawn to dusk and a running commentary made into a pocket tape recorder. The tape was then analyzed and the records for all observations are summarized in Table 9:3. Comparisons of the behaviour of the closely-followed deer with that of deer subjected to long-range observation showed that close following had little, if any, effect on their behaviour.

Hind A2 was pregnant on June 22nd when the first observation was made; she calved on June 30th and at all subsequent observations she was lactating to feed her calf. Hind B30 had calved on June 18th but her calf died immediately; she was thus not lactating throughout the observation period. The results show that on no occasion did the dry hind take as many bites as the lactating one. It is also apparent that fewer bites of grasses were taken by hind B30 than by hind A2. In absolute terms, hind B30 took fewer bites of ling than hind A2 in the first two periods, but subsequently took 20 to 50% more; in relative terms she always took a greater proportion of ling than A2 and she also took more blaeberry.

These observations cannot be converted into amounts of various species consumed; the amount of a plant consumed per bite may differ from plant to plant, in the same plant at different times and from animal to animal. They suggest, however, that the lactating hind selects a diet containing more grass and less heather than a dry hind.

Table 9:3 **The number of bites which two deer took from different plants between first and last light.* Supplementary food was available until July 10th.**

Hind	Date 1973	Total bites recorded $\times 10^3$	Percentage of bites from † Grasses	Ling	Blaeberry	Bracken	Polypody	Common Rush	Other
A2	June 22	12·3	77·3	17·5	5·1	0·5	0·1		Nil
	July 6	14·1	70·8	24·6	4·0	0·6	0·4	Nil	0·1
	July 20	18·2	83·1	10·6	1·2	0·2	0·4	0·6	3·9
	Aug. 4	18·2	82·7	13·7	0·3		0·1		3·4
B30	June 30	8·9	61·3	19·6	10·1	0·3	8·3	0·2	0·3
	July 15	6·6	67·2	25·1	5·6	Nil	1·9	0·2	Nil
	July 27	9·4	61·5	31·7	4·6	0·1	1·3	0·5	0·4
	Aug. 11	10·8	66·6	29·1	3·4	0·5	0·1	0·2	0·1

*The state of these hinds is given in Table 9:4; the observations were made on the same days as those of activity.
†Grasses included rushes and sedges other than the common soft rush *Juncus effusus*.

At the time these observations on herbage selection were made records were taken of the activities of the hinds and their duration. These are summarized in Table 9:4. They show that hind A2 (the lactating hind) grazed for much longer than hind B30 (the non-lactating one) and when A2 was pregnant she spent very little of her time simply moving about or standing; she preferred to lie. Hind B30 was remarkably consistent in the proportion of time which she spent lying, and this time always exceeded that spent lying by hind A2 when she was lactating. During daylight hours the deer did not appear to have favourite lying places. Figure 9:1 plots the positions adopted by 28 deer on seven days during the summer. There are some clusters but it was remarkable that the deer chose such a variety of places in which to lie. The activity patterns noted in the two deer are similar to those seen in other grazing ruminants.

One further point relates to patterns of defaecation. Unlike horses but like cattle and sheep, the faeces of the deer are dropped at random rather than in a defaecating area. It is not yet clear how far the deer has a behaviour pattern like that of cattle and horses which leads to avoidance of herbage in the close vicinity of areas dunged.

Figure 9:1 **The places in which 28 deer were noted to be lying on seven consecutive days during July. Note the absence of any great preferences.**

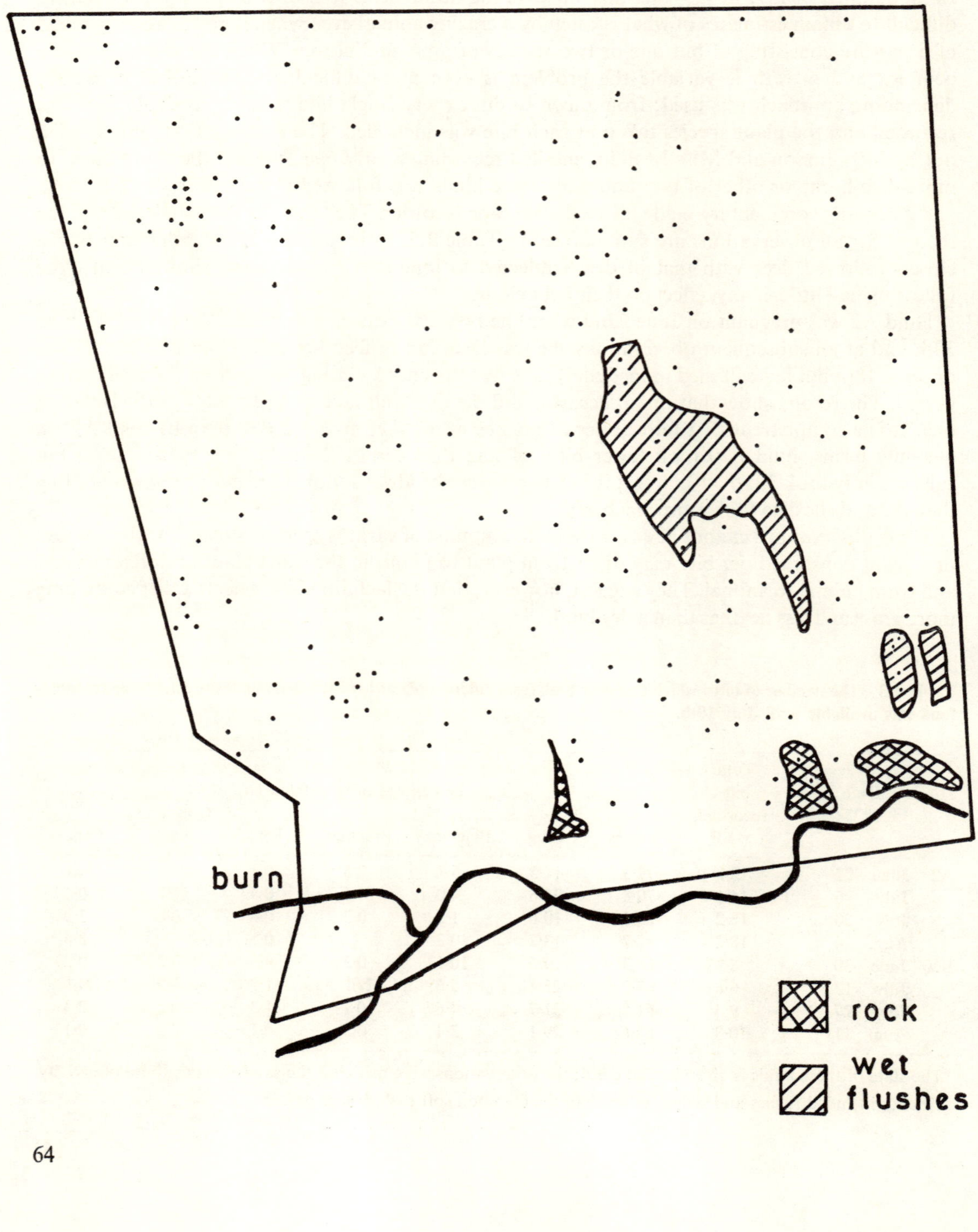

Table 9:4 **The activities of two hinds recorded between first and last light (24 h=1440 minutes)**

Hind	State	Date 1973		Time observed mins.	Percentage of time spent: Grazing	Lying	Moving or standing	Ruminating (included in lying and standing time)
A2	pregnant	June	23	1082	46·7	52·2	1·1	32·3
	lactating	July	6	1110	55·1	37·1	7·8	36·7
	lactating	July	20	926	63·1	18·8	18·1	22·0
	lactating	Aug.	4	987	63·8	30·1	6·1	27·6
B30	non-lactating	June	29	1110	38·2	44·3	17·4	17·2
	non-lactating	July	15	813	39·5	45·3	14·3	15·4
	non-lactating	July	27	1028	40·4	44·2	15·4	31·0
	non-lactating	Aug.	11	1015	39·1	42·9	18·0	25·5

Hind A2 calved on 30th June: supplementary feed available until 10th July.

Hind B30 calved on 18th June but her calf died immediately: supplementary feed available until 10th July.

The quality of the diet

A first measure of the quality of diet is its digestibility. The concentration of nitrogen in the dry faeces of a ruminant animal provides an index of the digestibility of the forage eaten, and has been employed in many studies of domestic ruminants grazing cultivated swards. The association of faecal nitrogen with quality of diet arises because most faecal nitrogen is of microbial origin. The amount excreted depends largely on the amount of microbial growth in the rumen and so on the quantity of fermentable material eaten. Most of the faecal dry matter, however, consists of plant remains, and the amount of this component excreted depends largely on the quantity of indigestible material eaten. Thus the higher the nitrogen concentration in faecal dry matter the better the quality of the diet.

To relate faecal nitrogen concentration to forage digestibility it is first necessary to measure the digestibility of the forage directly by feeding known quantities to animals and collecting all the faeces they void. Very little preliminary work has been done with deer. With sheep excellent relationships have been found when grass of a range of qualities has been offered and faecal-N concentrations can be used as a relatively accurate predictor of ingested herbage quality. When heather is fed Milne (1974) has found that the relationship is considerably less accurate and cannot be regarded as a suitable measure of the quality of herbage eaten by grazing animals. Since the proportion of heather in the diet is likely to vary seasonally it is to be anticipated that variation in values obtained may be great especially in winter.

But even so, it was thought that analysis of the faeces of the deer grazing at Glensaugh might show when the forage selected began to improve in digestibility in the spring and deteriorate in the autumn. Samples were taken directly from the rectum of deer at each monthly gathering. Six hind calves were chosen randomly in October 1971, and six yearlings, two hinds and four stags were added to this group in February 1972. In practice it was seldom possible to sample all the animals of the group and others were sampled instead to maintain a reasonably large number.

The results are shown in Figure 9:2. In these unsupplemented animals, depending on grazing alone, the digestibility of the forage eaten, as judged by faecal nitrogen concentration, began to improve in April. Forage quality reached a peak in June and July and then began to fall to reach a low level by November. Other animals receiving a supplement of concentrate cubes showed much higher faecal nitrogen concentration during the winter months, as would be expected, although in April 1972, when the supplement had been reduced to ½ lb per head daily, there was little difference between the groups (Table 9:5).

Many uncontrolled variables no doubt contributed to the scatter of values observed—the animals sampled each month varied, they were penned for some time before sampling, they were grazing different paddocks, some samples were very small and so poorly representative of the bulk of faeces voided, and so on. Moreover, the proportion of heather in the diet will have varied to some extent with season, and the crude protein of heather is rather poorly digested. Nevertheless, it is clear that there is a very marked difference in the quality of winter and summer herbage, and that for seven or eight months of the year the quality may be very poor when the main source of forage is likely to be heather.

Figure 9:2 **The mean concentration of N in samples of the faecal dry matter excreted by grazing deer. The vertical lines indicate the range of values.**

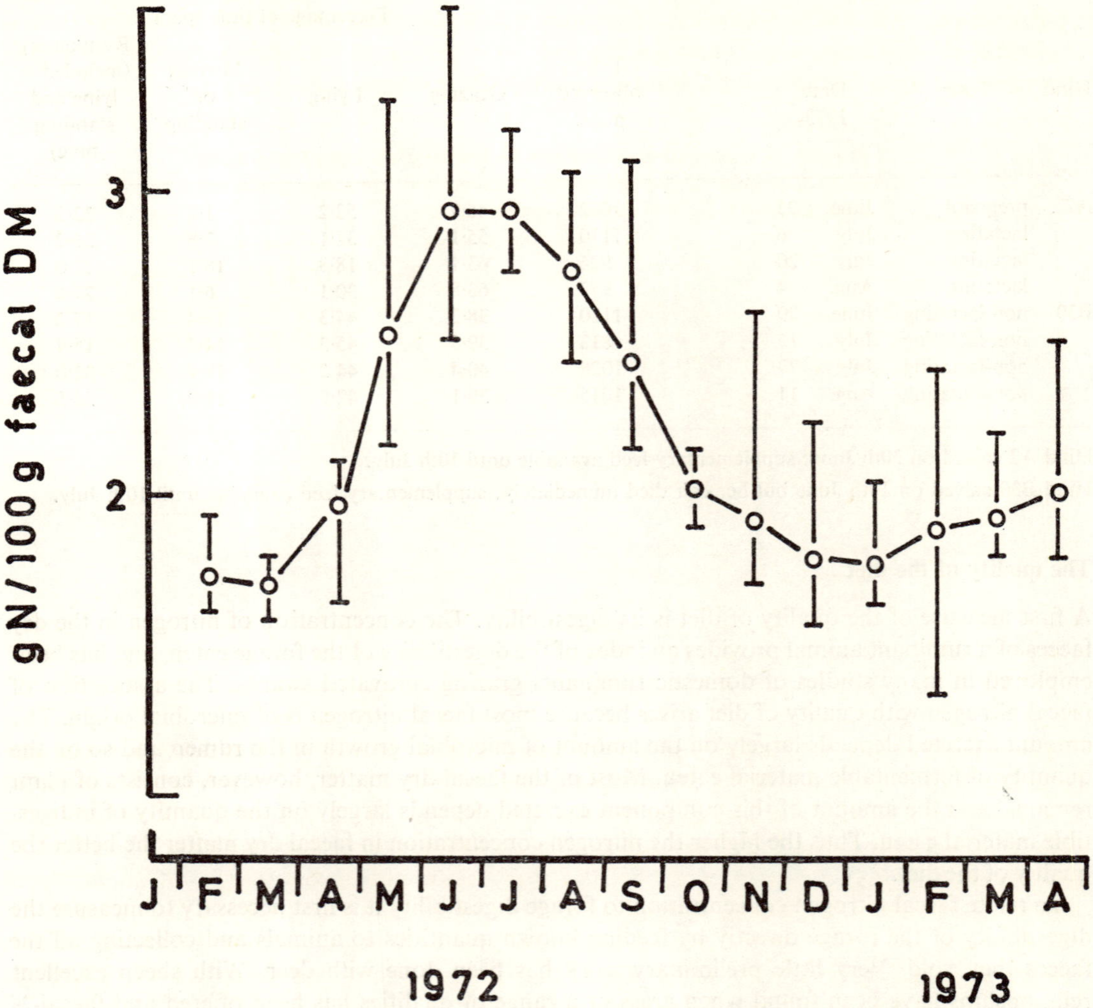

Table 9:5 compares the faecal N concentrations found during summer and winter at Glensaugh with values obtained at the Rowett using cut heather and grass given to deer, housed in metabolism cages, from which total faeces collections were made during periods of several days. Even though one should not make too close a comparison between the Glensaugh and Rowett results, it seems likely that the Glensaugh animals were succeeding in selecting a good-quality diet during the summer, and even in winter were finding better food to eat than the all-heather diets offered to the Rowett deer.

Tests of faecal samples as an index of rumen microbial activity in deer

Another way of finding the quality of the forage eaten by grazing animals is to record microbial digestive activity in samples of food residues taken from some part of the gut. Hobson and his colleagues (Hobson, 1970; Hobson and others, personal communication) took samples of rumen contents, intestinal contents and faeces from shot wild deer and examined rumen microbial activity, measured in various ways; this could be correlated with the nutritional status of the animals and the type of herbage being eaten. It was obviously desirable to develop alternative tests for samples which could be taken from living animals and which would allow one to take serial samples from the same animal. So tests on samples of faeces, similar to those made on rumen samples, were investigated.

In the first experiments on the rumen contents of shot deer the fermentation rates of the contents plus the substrates glucose (for general microbial activity), grass, heather, blaeberry and lichen (the first three being the main feeds of the deer), were measured *in vitro* and bacteria were counted in total and on different media. These tests, with estimates of ammonia and volatile fatty acid (VFA) concentrations, were also applied to samples of fresh faeces obtained from living deer. Whilst it was possible to take faeces from the rectum of small animals by hand, the only way found to sample large animals was by waiting for freshly-voided faeces. This limits the applicability of the method because whilst it was possible in winter in Glen Muick to stand amongst a stag herd, used to being given a supplementary feed at a certain time and place, and collect faeces, a similar collection would be difficult or impossible with a completely wild herd. At the deer farm it is possible to round up animals and confine them in a small area for faeces collection. For each experiment samples from about six deer were pooled and homogenized.

Table 9:5 **Concentration of nitrogen in faeces taken from free-grazing deer at Glensaugh by sampling from rectum, and from deer given cut forages at Rowett by total faecal collection**

	faecal N, g/100 g DM range	mean
Glensaugh deer (5–11 animals):		
Grazing+supplement, November 1971—March 1972	1·89—3·13	2·56
Grazing alone, June—August 1972	2·43—3·60	2·87
Grazing alone, November 1972—March 1973	1·33—2·60	1·86
Rowett deer (2 or 4 animals):		
Fresh heather, August 1968	1·32, 1·36	1·34
Dried heather, June 1969	1·44, 1·53	1·48
Poor-quality grass hay, 1968–9	1·37—1·65	1·51
Good-quality dried grass, 1968–9	2·18—3·72	2·87

The general results of 21 experiments on different deer or the same deer at different times may be briefly summarized as follows. In animals on normal grazing, fermentative activity in faeces was almost nil in winter and spring, but fermentative activity which appeared to be linked to the variety of herbage being eaten was measurable in summer and early autumn. The effect of supplementation was apparent, both in wild deer herds and at Glensaugh, in an increased activity dependent on the extent of supplementation. VFA and ammonia levels responded to the supplementation. Viable faecal bacteria tended to follow the same patterns. In young animals penned and feeding on starter rations faecal activity in all tests, except counts of cellulolytic bacteria (the starter ration was rich in starch), was very high.

The tests, then, could detect deer which were receiving adequate nutrition from summer pasture or supplementary feeding, and could possibly give some indication of herbage being eaten. However, they could not distinguish between a poor and a totally inadequate diet.

Sodium status of the deer

Sheep grazing pastures naturally deficient in sodium such as swards of timothy (Sinclair & Jones, 1968), and domestic and wild animals living in sodium-deficient high-rainfall regions such as the Snowy Mountains of Australia (Blair-West *et al.*, 1968), may be affected by sodium deficiency. As this may influence pasture selection and appetite, it was of interest to see if the deer grazing the peaty, well-leached pastures of the farm showed signs of this condition.

The ratio of sodium to potassium in mixed saliva is a sensitive index of the concentration of aldosterone in the blood and of sodium status, at least in sheep and cattle. At the monthly gatherings, therefore, mixed saliva was aspirated from between the molar teeth and cheek into a syringe, testing any deer which were anaesthetized with Immobilon for other purposes. For part of the year the stags had access to salt blocks containing urea and sodium chloride and were therefore likely to be well supplied with sodium. Other saliva samples were taken from stags housed at the Rowett Institute which received diets supplemented with sodium chloride. From time to time samples of vegetation taken from the more heavily grazed areas of the farm were also analyzed for their sodium and potassium content.

The results are presented in Tables 9:6 and 9:7 and in Figure 9:3. It can be seen that more than a third of the samples taken from animals that were not given any additional salt had Na^+/K^+ ratios suggestive of sodium depletion, and another quarter were borderline. The supposition that the ratio is as valid an index of sodium status in deer as in sheep was supported by the observation that the ratio was invariably high in animals receiving supplementary sodium. Low values were observed most often between July and November.

The sodium contents of heather, blaeberry, grasses and bracken were rather low at all times of year, but they were still an order of magnitude greater than those associated with chronic sodium depletion in Australian herbivores (Blair-West *et al.*, 1968). The rushes, which were grazed to a moderate extent, were a relatively rich source of sodium. Animals grazing even as little as 1 kg dry matter from this pasture daily would have an intake of about 30–50 m.equiv of sodium. In sheep, except during lactation, this would be quite enough to prevent the degree of depletion indicated by the salivary samples and so low sodium intake by itself does not appear to be sufficient to explain the condition.

Sodium depletion is usually associated with an increased appetite for salt or salty foods, and it may have served to make the urea/salt blocks more attractive. In any event, stags with access to the blocks evidently licked them enough to avoid any noticeable sign of depletion. Cattle with sodium deficiency commonly attempt to drink urine from their companions; hinds were sometimes observed to drink the urine of their sucking calves, but this behaviour was not noted in other circumstances.

Table 9:6 **The number of samples of mixed saliva in which Na^+/K^+ ratio was low (less than 5), medium (5–8) or high (more than 8). The samples were taken from deer anaesthetized with Immobilon. In sheep at least the low ratio would be diagnostic of sodium deficiency, the medium ratio of borderline cases, and the high ratio of sodium repletion.**

	Number of samples		
	Low Na/K	Medium Na/K	High Na/K
Glensaugh hinds	12	9	19
Glensaugh stags	12	6	6
Glensaugh stags with access to urea-salt block	0	0	13
Rowett stags receiving salt-supplemented diet	0	0	10

Table 9:7 **Sodium and potassium content of forages at Glensaugh (November 1972—October 1973)**

	No. of samples	m. equiv/100 g dry matter Na^+	K^+
Heather (*Calluna vulgaris*)	10	3·6	21
Blaeberry (*Vaccinium* spp.)	3	2·5	25
Mixed grasses	9	4·3	62
Rushes (*Juncus* spp.)	10	12·0	60
Bracken (*Pteridium aquilinum*)	3	3·2	39

Figure 9:3 **The ratio of Na^+ to K^+ in saliva from deer grazing at the farm and the effect of supplementation of sodium intake by provision of additional salt. Data for stags kept indoors at the Rowett Institute and given adequate sodium are included.**

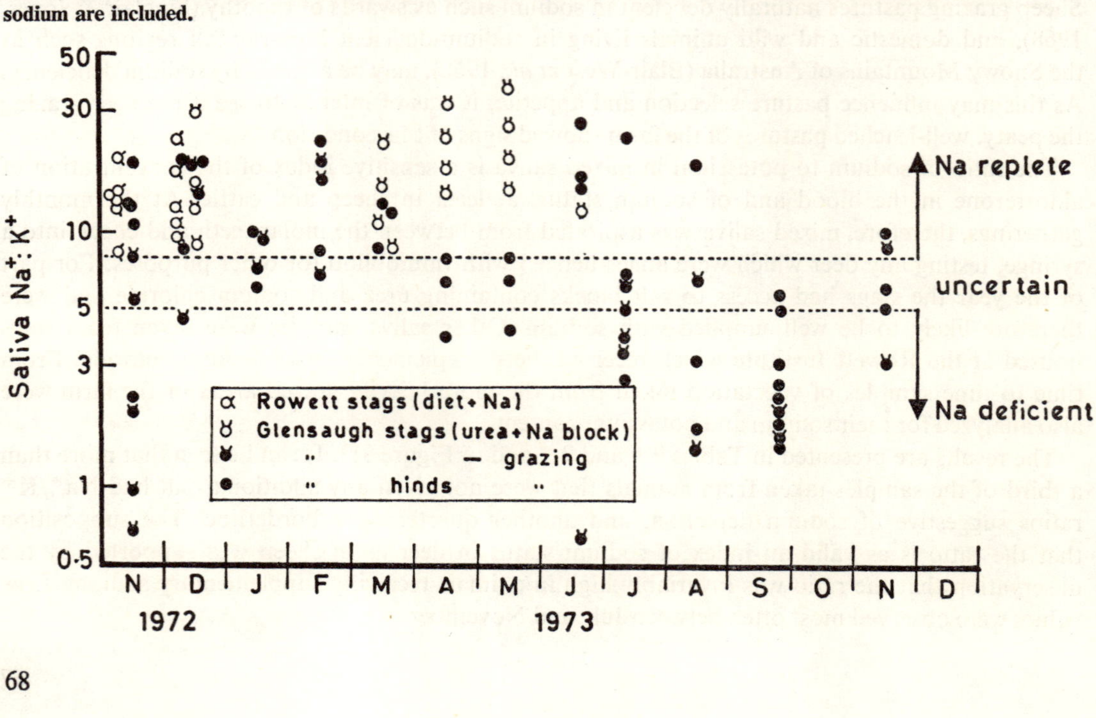

References

Blair-West, J. R., Coghlan, J. P., Denton, D. A., Nelson, J. F., Orchard, E., Scoggins, B. A., Wright, R. D., Myers, K. & Junqueira, C. L. (1968). Physiological, morphological and behavioural adaptation to a sodium deficient environment by wild native Australian and introduced species of animals. *Nature, Lond.* **217,** 922–928.

Hobson, P. N. (1970). Some field experiments on the rumen function of red deer, hill sheep and reindeer. *Deer,* **2,** 450–453.

Milne, J. A. (1974). The effects of season and age of stand on the nutritive value of heather (*Calluna vulgaris,* L. Hull) to sheep. *J. agric. Sci., Camb.* **83,** 281–288.

Sinclair, K. B. & Jones, D. I. H. (1968). Comparison of the weight gain and the composition of blood and saliva in sheep grazing timothy and ryegrass swards. *Br. J. Nutr.* **22,** 661–666.

Chapter 10. **Health**

It is clear from our experiences, recorded in Chapter 5, of rearing young calves caught in the wild and of the natural rearing of young calves by their mothers at the deer farm, that young deer calves are as susceptible to infections, to digestive upsets, and to the hazards of parturition, as are all other farmed stock. The number of deaths from these causes can clearly be influenced by the skill of the stockman in charge. A major objective of the study of the possibility of farming deer is to augment such husbandry and veterinary skills as have been derived from studies of the more usual farm species; much of this entails learning from our own mistakes.

Deer are thought to be susceptible to fewer diseases than sheep or cattle. This may be because of natural selection in ages past, when deer roamed free over the British Isles. It may also be because they are seldom exposed to some diseases, partly owing to their infrequent contact with farm stock. McDiarmid (1969) has stated that of 31 well recognized diseases, 27 occur in sheep, 19 in cattle, and only seven in deer, two of these being of no clinical significance. The diseases likely to be of importance in deer farming are parasitic pneumonia, liver fluke disease (in certain wet areas only), gastro-intestinal worms, tuberculosis, enzootic ataxia, which is a form of sway-back occurring in park deer in England, and, if deer come into close contact with infected cattle, possibly brucellosis. In addition, deer are known to be seriously troubled by warble fly.

As large areas of the rough grazing land which may prove to be suitable for the farming of deer are infested with ticks, mainly with *Ixodes ricinus*, the apparent resistance of deer to the major tick-borne diseases is an asset to the potential deer farmer. In addition, resistance to the clostridial diseases, such as blackquarter, enterotoxaemia, and black disease, and to 'strike' by the various blowflies may mean that very few gatherings for disease control and prevention will be needed in the course of a year.

Tuberculosis

In mammals and birds infection with *Mycobacterium tuberculosis* usually causes a slowly progressive loss of condition resulting ultimately in emaciation and death. Some of the abdominal organs from an emaciated stag found on the point of death in the wild, were sent to the Rowett Institute where caseating, purulent lesions, typical of tuberculosis in other species, were found in its mesenteric lymph nodes. This pus contained numerous acid-fast bacteria, which on culture had the characteristics of *M. tuberculosis var. avium* (*M. avium*). This is the variety which has on several other occasions been isolated in Scotland and England from shot wild deer (McDiarmid, 1969).

At the Rowett Institute in 1970, 12 young calves were obtained from a sporting estate for an experiment designed to show their growth potential. They were penned individually and after being bottle-reared on cows' milk, were weaned gradually on to a barley-based diet which was offered *ad libitum*. Until they were 13 weeks old all grew well and at a comparable rate, but during the succeeding four weeks one ceased to grow, its appetite diminished and it developed intermittent diarrhoea. In its 17th week, the diarrhoea became profuse and blood-stained, and it died. Examination *post mortem* revealed a gross enlargement of all the lymphatic tissue within the abdomen including the Peyer's patches, with an extensive, ulcerative inflammation of the intestines. Microsections were later found to contain numerous acid-fast bacteria. Unfortunately at the time tuberculosis was not considered as a possible cause and specific cultural methods to isolate *M. tuberculosis* were not used. Subsequent events left little doubt that this had indeed been a case of tuberculosis, remarkable for the speed with which the infection had overwhelmed the deer's defences. At the time the five pens which formed the wing of the building in which the sick deer had been housed were scrubbed thoroughly and fumigated with formaldehyde. A year later a second of these four calves, which had been transferred to a zoo six months previously, became emaciated and died. Post-morten examination revealed caseating lesions typical of tuberculosis and *M. avium* was isolated. It is possible, of course, that the animal had acquired infection after leaving the Institute.

The comparative intra-dermal tuberculin test, using Weybridge purified protein derivative avian and mammalian tuberculin, was carried out on deer remaining in the pens at the Institute by veterinary officers of the Animal Health Division on several occasions with the results shown in Table 10:1. The tuberculin injections were made intra-dermally in the middle third of the side of the neck and skin thicknesses were measured initially and at 72 h, that is exactly as in the official British test of cattle with the avian site above the mammalian site and the sites separated by about 6 cm. Reaction to the avian tuberculin occurred in some deer. The skin thicknesses measured at 72 h at the site of the avian tuberculin injection exceeded the initial measurement by at least 3 mm. Animals which reacted were retained until their experimental purpose had been achieved. Several were tested on two or more occasions and these gave a reaction each time to avian tuberculin. On only one occasion was there a reaction to the mammalian tuberculin but on retest there was no reaction.

Once there was this confirmed diagnosis of tuberculosis, all deer at the deer farm were tested on two occasions by veterinary officers of the Animal Health Division. Only one reactor was found. This was an avian reaction in a hind which had passed the tuberculin test at the Rowett Institute four months previously and had been transferred to the farm. It was slaughtered immediately thereafter.

Four reactor deer were slaughtered for carcase analysis and the opportunity was taken to search for lesions of tuberculosis. All were in good condition at the time of slaughter. On only one occasion was a purulent lesion found. This was a spherical abscess in a retro-pharyngeal lymph node. Cultures of this abscess gave growth of a diphtheroid bacterium and also of *M. avium.*

Table 10:1 **Comparative intra-dermal tuberculin tests of deer penned indoors at the Rowett Research Institute**

Date	8.11.71	10.1.72	18.9.72	12.12.72	31.8.73	26.10.73
No. tested	20	15	19	16	12	6
No. negative	17	11	16	14	9	5
Reactors	Binti (A)	Binti (A)k				
	Asali (A)	Asali (A)	Asali (A)k*			
	Jitu (A)	Jitu (A)	Jitu (A)	Jitu (A)	Jitu (A)k	
		Inzi (A)t				
A = *Avian*			Hamisi (A)k			
M = *Mammalian*				2205 (A)	2205 (A)	2205 (A)
k = *killed*					S13 (M)	[S13 negative]
t = *transferred*						

**M. avium* isolated from retro-pharyngeal lymph node

The above results confirm the susceptibility of the red deer to infection with *M. avium.* The circumstantial evidence from the first death suggests that *M. tuberculosis* can become a fulminating infection in a comparatively young calf, causing death as early as 17 weeks of age. It can also cause a debilitating illness in deer as in other species. The comparative intra-dermal tuberculin test can, however, be applied and is readily interpretable. From only one reactor out of four was it possible to find a lesion and to isolate *M. avium.* This individual had reacted on three occasions and during the 10 months that had elapsed since it had first reacted to the test the disease had not progressed to a stage beyond causing a small abscess in a lymph node. It is probable that the other reactors had even smaller lesions; they showed no detectable sign of disease other than the positive reaction to the test.

An unsolved mystery is how the infection continued to spread among the deer housed indoors. Undoubtedly the first sick animal presumed to have had tuberculosis voided very large numbers of mycobacteria into the environment during the few weeks of 1970 when it had intermittent diarrhoea. None other of the subsequent reactors was an 'open' case. The source of subsequent infection may have been the sick calf in 1970, in spite of our attempts to disinfect. Alternatively, wild birds could have contaminated the deer's food before or after its harvest.

It is interesting to speculate on the mode of transfer of tuberculosis in the wild. Wild birds frequently harbour the avian infection and conceivably could contaminate the food of deer. On deer farms it is likely that birds will share food offered in troughs to deer and by contaminating the food with droppings they could possibly transmit the disease to the stock.

Examinations of serum for antibodies to known diseases

Brucellosis. Deer are known to be susceptible to infection with *Brucella abortus*. In Russia serological reactions and isolations of the organism have been obtained from the Maral and Altai deer (*Cervus elaphus maral* and *altaicus*). These deer may have lived at one time in close contact with infected cattle. Many of the American elk or wapiti (*Cervus canadensis*) of Yellowstone National Park were found during the planned reductions of numbers in 1962 to have antibody titres in their blood serum which classed them as reactors under the U.S. brucellosis eradication scheme. The Yellowstone elk herds lived in close association with the Park bison which were also found to be infected. In Britain McDiarmid (personal communication) has examined thousands of serum samples from red deer in Scotland, England and Wales but has not yet found any with a detectable level of antibody.

At the deer farm blood was taken from all deer present in September 1972 and November 1973. There was no reactor.

Leptospirosis is a disease of man and animals caused by one of many species of bacteria belonging to the genus *Leptospira*. These bacteria can live outside the animal body long enough for transmission to take place by down-stream transfer in drinking water. Rodents are important reservoirs of infection and can transmit disease to farm animals by contaminating their food with urine. The disease is characterized by jaundice and anaemia.

Twigg & McDiarmid (1973) examined the sera of 345 deer from various parts of Britain, including 210 red deer from the Isle of Rhum, Perthshire and Middlesex for antibodies to 15 species of *Leptospira*. They found very low levels of antibody in only two.

In October 1972 serum samples from 100 healthy deer at the deer farm were examined in the Department of Veterinary Pathology at the University of Glasgow Veterinary School, for antibodies to 16 species of *Leptospira* by Schüffner's test using living cultures of *L. interrogans*. Two sero-reactors were found—a stag with a titre of 1:300 to *L. ballum* and a yearling hind with 1:30 to *L. icterohaemorrhagiae*. Two months later the stag's serum titre was still 1:300 but the hind's had become negative. Five months after the first test the titre was still 1:300 but after a year it also had become negative. Although the stag's titre persisted for at least five months, there was no evidence of any associated illness. The hind calved normally at two years old in July 1973.

Louping ill is a tick-borne disease caused by an arbovirus which in sheep and cattle localizes in the brain. The resulting encephalitis causes mental derangement; the affected animal leaps in the air (loups) and becomes unable to feed. Antibody to louping ill virus has been found in deer but the illness, which is very characteristic and would be readily identified by deer stalkers or naturalists, has never been reported.

Serum samples from 60 calves on the deer farm were taken in October 1971. The calves were from estates in many different parts of Scotland and were approximately four months old. Two of the samples were found at the Moredun Institute, Edinburgh, to have low titres (1:10 and 1:20) to the louping ill haemagglutination inhibition test. Sera from six young stags were all negative. A month later the titres of the two calves had declined to zero and 1:10. It is most likely that the titres represented the residues of maternal antibody acquired from colostrum.

Sheep and cattle develop clinical disease if their antibody levels are low at the time when they first become infested with infected ticks. This can happen in young animals if they are born some time before the ticks become active or if their dams are removed to tick-free pasture to lamb or calve and they are then returned to the tick-infested pasture at a time when the ticks are active. The apparent absence of clinical disease in wild deer may indicate either that the central nervous system of the red deer is inherently resistant to infection with the virus, or more probably, that because red deer calve in June and July after the ticks have become active, the dams are able to give their calves adequate protection via colostrum. A system of deer farming could be devised in which the deer would be removed to tick-free pasture for calving and would then be transferred back to tick-infested pasture with their young calves. This could be dangerous. Vaccination against louping ill might be necessary in such a circumstance.

Toxoplasmosis is the disease caused by the protozoal parasite *Toxoplasma gondii*. Infection of sheep can cause abortion. In man it can cause damage to the inner structures of the eye. Generalized infections can result in death. Transmission may be congenital. The domestic cat can harbour the infection in oocyst form in its intestines and as a carrier may indirectly spread infection to other mammals. There are undoubtedly other means of spread but these are not yet known. It is not known whether *T. gondii* causes disease in deer.

In September 1972, 98 serum samples from the deer on the deer farm were examined at Raigmore Hospital, Inverness, by the haemagglutination test for *T. gondii* (Thorburn & Williams, 1972). One young stag had a high titre (1:128). One yearling hind and one female calf had titres of 1:32 and 11 deer had titres of 1:16. The remainder were 1:8 or less. This suggests that 14 deer had had contact with the parasite at some time. The titre of the stag's serum had declined five months later to 1:32 but a year later had again risen to 1:128.

Sheep kept indoors at the Rowett Institute are at high risk from this infection (Sharman *et al.*, 1972). Serum from three stags and two castrates, also housed at the Institute, were tested in February 1973. Two of the stags had titres of 1:16 which may indicate contact with infection. No illness had been noticed.

Respiratory infections. In the course of routine blood examinations blood from a stag was found to have a high erythrocyte sedimentation rate. The sample was sent to the Bacteriology Department of Aberdeen Medical School where it was subjected to complement fixation (CF) tests for influenza A & B, parainfluenza, adenovirus, the syncytial respiratory virus, psittacosis and Q fever, all of which were negative. It is interesting that to the CF test for the mycoplasma of the human respiratory tract *M. pneumoniae* it had a titre of 1:64. This is an isolated observation the significance of which is not known.

Babesia. On certain tick-infested pastures in Scotland cattle become infested with the blood-borne protozoal parasite *B. divergens* as the result of tick bites. Most animals carry the infection for life without ill effect, but individuals under conditions of stress not yet defined, become ill with an active parasitaemia and haemoglobin in their urine. The disease is aptly named redwater. During the illness the parasites invade a relatively large proportion of the erythrocytes (red blood cells) and can be readily seen in stained blood films.

Deer also live on these pastures. It is not known whether they too develop redwater. Haemoglobinuria would be difficult to detect in deer in the wild. It is known, however, that some red deer in certain areas of Scotland have antibody to *B. divergens* in their blood (Latif & Adam, 1973).

Sera taken from 122 deer at the deer farm were tested in the Zoology Department of Edinburgh University with an indirect fluorescent antibody test using *B. divergens* as antigen. Seven were positive and when related to their source of origin gave useful information about the extent of infection in Scotland.

Parasites

(*a*) *Gastro-intestinal helminths.* Faecal samples were collected at each monthly gathering from selected individuals and were examined for worm eggs by the McMaster technique, which is the test generally used in diagnostic laboratories. The highest egg count ever recorded was 300/g which is quite low. Most samples had counts of 100/g or less. The Baermann technique was also used on samples that had been stored to allow eggs to hatch. The technique detects very small numbers of larvae. Results indicated that most deer had some nematode helminths in their gastro-intestinal tracts. There was no indication that there was any increase in the helminth burden in successive years even with the increase in the numbers of deer in the years under review.

(*b*) *The head fly* (*Hydrotaea irritans*). This fly is all too familiar to people who walk through bracken or heather which grows close to woods. It resembles the housefly. Clouds of many hundreds swarm around the heads of hikers on hot sunny days. The deer are similarly pestered, but particularly the stags when their antlers are in velvet. The flies land on the velvet and as the deer have no effective means of dislodging them from the antlers which at this stage are very sensitive, on bad days they spend most of their time trying to escape and consequently spend little time grazing. July and August are the months when the flies are most active.

In preliminary studies of the problem a means of attracting the flies with CO_2 was developed: this permitted identification. Remarkably little appears to be known of the fly's biology. Attempts were therefore made to discover where it over-wintered, by setting out emergence traps in various parts of the farm, but with no success.

In early August both male and female flies were observed but later in August only females were found. When trapped and taken to the laboratory these laid eggs from which second instar larvae emerged and grew to 0·5 cm. Unfortunately at this point studies had to be discontinued but the intention is to resume them when the next opportunity arises, as pestering of the deer by this fly may well be a major hazard of deer farming and a means of control must be sought.

(*c*) *Warbles.* These are the lumps that develop on the backs of deer and cattle as the larvae of the warble fly grow beneath the skin. The deer warble fly is called *Hypoderma diana.* It affects both red and roe deer but not cattle. Cattle warble flies do not affect deer.

The deer warbles can first be detected in January. After arriving in the deer's back each larva bores a small hole in the skin and slowly enlarges beneath the skin. In April, when it has grown to about 2 cm long × ½ cm in diameter, the larva, by now mature, squeezes to the exterior through the hole and falls to the ground, where it pupates. In summer, an adult fly emerges from the pupa and takes wing. When it finds a deer it hovers and darts, laying its eggs on the hairs. The fly then dies, but larvae hatch from the eggs and burrow through the skin into the deer's body. The cycle is complete when the larvae find their way to the backs in January.

Some red deer calves that were caught in the wild in June 1970 when they were only a few days old, were reared indoors at the Rowett Institute where there are no warble flies. Warbles appeared in their backs in spring 1971. From this observation it was concluded that *H. diana* can lay its eggs on calves soon after they are born in June.

Warble flies can be controlled in cattle by pouring an organophosphorus preparation on to the backs. It is absorbed through the skin, to give a blood and tissue fluid concentration high enough to kill the warble larvae. In February 1972 we applied this preparation to the backs of deer which had warbles. In March we found only dead warbles in these animals. In the next warble season we tested the effect of pouring the preparation on to the backs before the warbles could be detected. We treated half of the yearling hinds and calves in December 1972 and re-examined them in February and March 1973. We found that fewer of the treated deer had warbles than the untreated deer (Table 10:2).

In addition to showing the beneficial effect of prophylactic treatment, Table 10:2 shows that a smaller proportion of calves than yearlings were affected. This may either be because warble flies were not active on some of the estates from which the calves came, or because much of the fly egg-laying activity took place before the calves were born.

(*d*) *Ticks, lice and keds.* Several of the calves which were caught on estates when only a few days old, were heavily infested with unengorged ticks (*Ixodes ricinus*). Treatment was not applied and the ticks, some of which were subsequently seen to be engorging with blood, presumably dropped off during the period when the calves were being fed indoors. Although death from infections did take place during the early rearing periods (Chapter 5) and some of these could possibly be attributed to the introduction of infection through the skin by the bites of ticks, this seemed not to be the case. In lambs such infection usually results in chronic lameness associated with the formation of pus in the joints or in multiple abscessation of the liver (tick pyaemia). Lesions of this kind have not been reported by observers of wild deer. The infecting bacterium in lambs is a haemolytic staphylococcus, which was not found in the deer calves which died at the deer farm.

Table 10:2 The effect of treatment with organophosphorus 'pour-on' in December on the subsequent development of warbles

	Treated	*Not Treated*
No. of yearling hinds with warbles	5/23	17/26
No. of calves with warbles	2/13	1/13

The area of the deer farm was thought not to carry ticks in the days when sheep grazed there. In the three years of deer farming ticks have not been found at the monthly gatherings on any of the deer.

Lice were present on the ears of several calves when they were first caught. Treatment was not given. At the deer farm louse infestation did not subsequently occur. Specimens were identified as a species of the biting louse (*Trichodectes*).

At the deer farm a single specimen of a ked was found on each of two occasions from two different deer. The keds were removed and were identified as *Lipoptena cervi* which during part of its life cycle has wings.

(*e*) *Lungworms.* Clinical signs of hoose or husk, the parasitic bronchitis caused by *Dictyocaulus* lungworms, have never been detected at the deer farm. An additional reason for discounting parasitic bronchitis as a hazard at the present stage of development of the farm is that routine examination of faeces collected at the monthly deer gatherings has consistently failed to reveal any lungworm larvae. However, during the *post mortem* examination of a hind 1 year 5 months old, which died when under the influence of Immobilon (Chapter 8), a few adult lungworms associated with mild localized bronchitis were found. The worms were indistinguishable in their morphology from *D. viviparus*, the bovine hoose worm. As the hind had been reared as a calf in a pen earlier occupied by cattle, it seemed possible that the infestation could have been acquired from contamination of the pen with cattle faeces. The susceptibility of red deer calves to *D. viviparus* of bovine origin was later tested experimentally. Subsequent routine examination of faeces from the other deer on the farm has not so far yielded any lungworm larvae.

On several occasions red deer lungs sent from sporting estates in Scotland were found to contain circular raised lesions beneath the pleura which contained eggs and larvae of the lungworm *Bicaulus sagittatus*. Although similar larvae were found in the faeces of two hinds which came from Rhum and were kept indoors at the Institute for about six months for lactation studies, the larvae were not found in their faeces after the hinds were transferred to the deer farm. In addition to the above lesions, two hinds from one estate had extensive lesions of pneumonia.

In contrast with infestation with *Dictyocaulus*, where young animals are more severely parasitized than older animals, which develop an immunity, infestation with *Bicaulus* appears to be commoner in older deer than in young.

It is important for the future development of deer farming to know whether the parasites of sheep and cattle, which are likely to share pastures with deer, are a hazard to the deer. As the lungworms found in a hind at Glensaugh were indistinguishable in their morphology from *Dictyocaulus viviparus*, the bovine lungworm, and as the hind had, as a very young calf, been in indirect contact with cattle, it was considered important to test the susceptibility of deer to the cattle lungworm. An experiment was therefore carried out.

Infective larvae obtained from cattle were kindly supplied by Allen & Hanburys Ltd. These were given in turn to two deer calves kept in pens at the Rowett Institute. The larvae were given as single doses by the natural route, namely by mouth. The calves were examined clinically each day thereafter and faecal samples were examined daily to discover whether the life cycle had been completed within the deer.

The first calf was given 88 larvae/kg liveweight when it weighed 28 kg. On the 28th day after dosing, larvae appeared in the faeces and were found in relatively small numbers each day until day 33. There was no coughing or other sign of illness. The dose given and the time interval between dosing and the recovery of larvae accorded with the time required in bovine calves for establishment of worms in the lung which on reaching maturity would lay eggs that would subsequently be recoverable as larvae from the faeces.

The second calf was then given twice as many larvae (related to body weight). Much larger numbers of larvae appeared in the faeces, starting on day 21 and continuing until day 51 when the calf was killed. This calf had coughed when disturbed but otherwise had not become ill. On *post mortem* examination there was a moderate degree of broncho-pneumonia and adult *D. viviparus* were found in the bronchi and trachea.

This experiment indicates that deer calves are susceptible to infection with the cattle lungworm.

The first calf was then re-dosed 3 months after its first experimental infection, with three times its original dose, again related to body weight. This time larvae did not appear in the faeces. Unfortunately there was no control calf available for this experiment but it is tempting to assume that as happens in cattle, some measure of immunity had resulted from the first dosing.

(*f*) *The nasal botfly* (*Cephanomyia auribarbis*). A hind which was taken from the deer farm in February 1974 for slaughter at the Rowett Institute, was found to have eight larvae, morphologically resembling *Oestrus ovis*, the sheep bot, in its pharynx. The hind had not shown any sign of discomfort. The larvae measured about 10 mm × 3 mm and had their mouth parts buried in the tonsillar crypts. There was no inflammatory reaction. In August flies like small bumble bees were caught as they zoomed close to the deer. They were identified at the North of Scotland College of Agriculture as *C. auribarbis*. This fly is viviparous, that is it lays live larvae not eggs. This it does in summer by hovering close to the nostrils and laying small larvae one after the other in the nostrils. The larvae then migrate up the nose to the pharynx or frontal sinuses where they enlarge by growing and moulting. In the following spring the larvae descend the nostrils, fall to the ground and pupate. From the pupae the adult flies emerge in summer.

It is of interest that the hind had been treated for warbles two weeks previously with an organophosphorus preparation, poured on to the back. Although this drug is absorbed through the skin of cattle and the resulting level in the blood and tissue fluids is high enough to kill migrating warble larvae, it clearly had little or no effect on these nasal bots.

Accidents on the hill

A *fore-limb fracture* was sustained by a young stag which had been born indoors at the Rowett Institute and suckled there by its dam. This staggie had grown rapidly and weighed 83 kg when it was transferred at 40 weeks old to the deer farm. Two months later it was seen to be very lame. Although its left fore-limb was swinging limply, it had no difficulty in walking and cantering. It was loaded into a trailer, supported by straw bales, and driven to the Institute where the fractured limb was radiographed with the deer narcotized with Immobilon.

The injury was found to be a compound, compacted, comminuted fracture of the mid shaft of the radius-ulna. Reduction of the fracture by traction was not attempted because of the forces that would be involved afterwards. The limb was immobilized with plaster of Paris bandages and the deer was confined to a small pen. Six weeks later a bridging callus was seen on radiographs to have incorporated both radius and ulna. The cast was removed two weeks later and the deer soon regained full use of the limb.

This fracture occurred on rough hill land in an animal which had not seen such territory until it was a year old. Moreover it received no guidance from its mother or from any other deer because, as an incomer, it was not accepted by the other social groups of deer which had already been formed. Evidently it paid for its inexperience with a broken limb.

Congenital abnormalities

The brains of young calves which died were examined by freezing the head and then bisecting it by sawing in a longitudinal and vertical plane. The shape of the cut cerebellum varied. In most it was almost circular—like a square with rounded corners. In a few others it was foreshortened, as if compressed into a crescent by pressure from the posterior portion of the skull. Of course if this is a true congenital defect and not an artefact the skull would take its shape from the underlying brain as it developed and the results would not really be compression of the brain by the overlying bone. A similar defect has been reported in bovine calves; its significance is unknown.

A 'parrot-mouthed' female calf was caught and taken to the deer farm. At the end of its first winter when it had been offered concentrate pellets in a trough along with the others it was in quite good condition. It continued to grow during the summer when the yearlings were fending for themselves but its weight and condition were poor compared with the others.

One calf was born with flexed back tendons so that when it stood up it could not put the soles of its hind feet on the ground. Instead it hobbled about taking its weight on the skin of its flexed pasterns. The fore-limbs were normal. The limbs were gently but firmly straightened and splints of corrugated cardboard and wood were applied with string so that when the calf got to its feet it was standing on tip-toe. The splints were carefully arranged and were re-adjusted when necessary to ensure that undue pressure was not being exerted on the soft tissues, otherwise the blood supply might have been shut off and gangrene would have developed. Gradually the tendons stretched until it became possible to remove the splints permanently without the pastern joints resuming their flexed position. The calf became normally agile and grew well. The primary lesions appeared to be in the hocks. When the calf was born it could be seen that the points of both hocks had been rotated upwards and forwards in the uterus and the bones had ossified in this position. Now that the animal is a mature hind this abnormal conformation is still evident. As we think this is not a heritable characteristic but rather an abnormal positioning of the calf as the bones developed inside the uterus, the hind has been mated.

References

LATIF, B. M. A. & ADAM, KATHERINE M. G. (1973). Antibody to *Babesia* in Scottish red deer (*Cervus elaphus*). *Nature, Lond.* **241**, 476–477.

McDIARMID, A. (1969). Diseases of red deer and their possible significance in deer farming projects, veterinary preventive medicine and public health. In *The Husbanding of Red Deer*, pp. 50–54. Ed. by Bannerman, M. M. & Blaxter, K. L. (See references to Chapter 1).

SHARMAN, G. A. M., WILLIAMS, KATHLEEN A. B., THORBURN, H. & WILLIAMS, H. (1972). Studies of serological reactions in ovine toxoplasmosis encountered in intensively-bred sheep. *Vet. Rec.* **91**, 670–675.

THORBURN, H. & WILLIAMS, H. (1972). A stable haemagglutinating antigen for detecting toxoplasma antibodies. *J. clin. Path.* **25**, 762–767.

TWIGG, G. I. & McDIARMID, A., (1973). The low incidence of leptospirosis in British deer. *Vet. Rec.* **93**, 98–100.

Chapter 11. **Intensive investigations**

Before the farming enterprise began in 1970 a small programme of work had been in progress at the Rowett Research Institute in which the red deer was the experimental animal. Much of this work was concerned with the comparative physiology of the deer and particularly with any differences between its digestion and metabolism and that of the domesticated ruminants. They were intended as much to deepen our understanding of sheep and cattle, by showing under carefully controlled conditions how these long-domesticated animals differed from a wild ruminant, as to examine the deer.

With the advent of the deer farm, the skills that had been acquired in handling these animals in close confinement for experimental purposes were made use of by partially re-orientating what is still a relatively small programme towards answering questions more immediately relevant to the farming operation. Some of these questions were relatively simple ones. What is the quantity and composition of hinds' milk? How fast can a well-fed deer calf grow, and if it grows very quickly, will it deposit a great deal of fat? What is the quality of a deer carcase from a butcher's point of view? Is any advantage to be gained by castration? Other questions related to the major problem of any farming operation, namely, how to match feed resources to animal nutrient needs. What are the physiological demands made by lactation and pregnancy and what are the protein and energy requirements of deer during growth and lactation? How well adapted is a deer to the intensely seasonal nature of its natural environment and pasture?

Partial answers to many of these questions have now been obtained but much remains to be done. In the course of this work, and with the continuing stimulus provided by observation of animals at the farm, entirely new questions have arisen. For example, experimental observations suggest that the low intake of feed and consequent low growth rate of deer in winter is not wholly due to poor quality of feed but to an innate rhythm in appetite related to the annual light cycle; this not only requires confirmation but raises problems related to cold resistance in winter, for resistance to cold is markedly reduced by undernutrition.

In this Chapter these intensive investigations, which have been or will be published elsewhere, are briefly summarized.

Digestion and metabolism

The first experiments were undertaken in 1966 by G. M. O. Maloiy, a Kenyan then studying for his Ph.D. Two mature hinds and two ewes were provided with rumen cannulas and given identical roughage rations while confined in metabolism crates. It was found that fermentative digestion in the rumen was virtually the same in each species, yielding very similar concentrations of volatile fatty acids and ammonia, at a similar pH, and leading to overall apparent digestibilities of organic matter and nitrogen that differed only a little (Maloiy & Kay, 1971). Closer study indicated that the time food was retained in the digestive tract seemed to be somewhat shorter in deer than in sheep. In one experiment on two deer and two sheep, hay or dried grass was rationed approximately according to metabolic body weight; the mean apparent digestibilities of the dry matter of the hay and grass respectively were 60 and 73% for the deer and 64 and 77% for the sheep, and the mean retention times of stained hay and grass were 64 and 58 h in the deer and 95 and 83 h in the sheep. Moreover, the voluntary intake of these roughages, when they were supplied *ad libitum*, was usually greater by the deer (Kay & Goodall, unpublished). Figure 11:1 illustrates one trial of the series. The animals were given their ration of hay, which included a small amount of stained hay, at zero time. Thereafter all faeces excreted were weighed, sampled and examined for their content of stained particles so that a measure could be obtained of the time food residues took to traverse the digestive tract. The greater the intake the shorter the retention time, but independently of this the deer retained the food residues for a shorter time than the sheep.

Similar differences in the digestibility of dry matter were observed when rations of pelleted cereals plus straw were given to deer and sheep (Maloiy, Kay, Goodall & Topps, 1970); the slowly-digested cellulose was 11 percentage units less well digested by the deer. It is interesting that subsequent experiments in which the ability of growing deer calves and lambs to digest more rapidly-digested concentrate rations was examined did not reveal any marked difference between the species (Pollock, Webster & Smith, unpublished). One may conclude tentatively that the propulsive activity of the digestive tract of the deer studied was somewhat greater than that of the sheep leading to more rapid passage of food residues and so to a slightly reduced efficiency of digestion of slowly digested roughages, but to considerably greater appetite. This could be a net advantage to a species which has abundant food of poor quality available to it throughout most of the year.

Figure 11:1 **The recovery of stained hay given at zero time in the faeces of deer and sheep. Deer excreted the hay more quickly than did sheep**

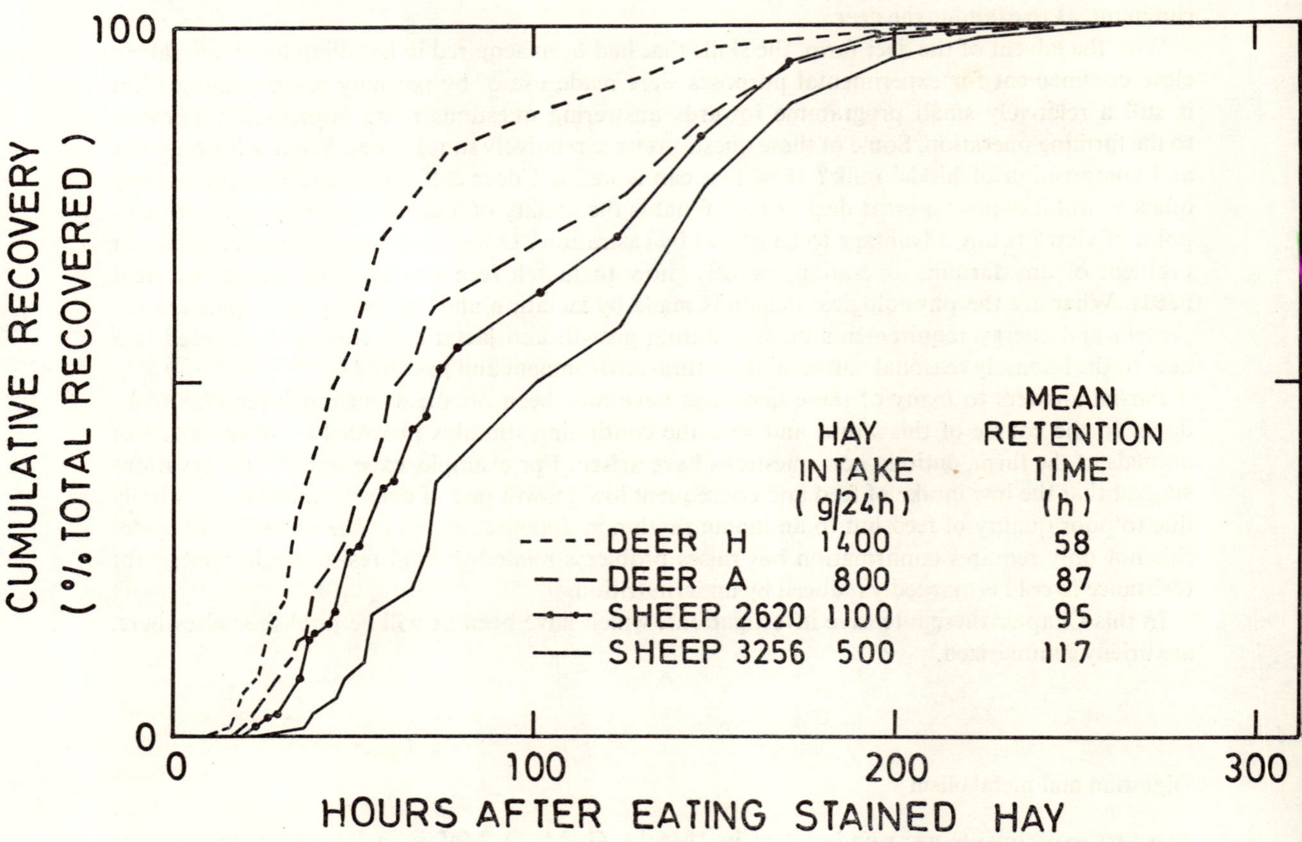

A more detailed study of nitrogen metabolism (Maloiy *et al.*, 1970) showed that deer and sheep digested the nitrogen-containing materials of cereal-straw mixtures equally well and retained similar amounts of nitrogen in their bodies, whether the food was rich or poor in protein and whether water was restricted or not. This suggests that deer have no more than the usual ability of ruminants to convert dietary nitrogen and endogenous urea into digestible microbial protein. However, deer differed from sheep in excreting relatively small quantities of allantoin and uric acid in their urine. This was rather puzzling, since in ruminants allantoin and uric acid are mainly derived from microbial nucleic acids and the latter, like microbial protein, appear to be synthesized as readily in deer as in sheep (Razzaque, Topps, Goodall & Kay, 1973). Presumably intestinal digestion or absorption and metabolism of nucleic acids follows a rather different course in the two ruminants.

In further studies, Maloiy (1968) compared the ability of deer and sheep to adapt to a restricted water intake. Sheep, originating in seasonally arid areas of Eurasia, withstand considerable dehydration before their appetite begins to fail. Deer, which are natives of temperate woodland, not surprisingly showed much less ability to contend with dehydration; their appetites were affected at an earlier stage, their faeces remained more moist and their urine less concentrated than in the sheep. Their voluntary water intake is at least 50% greater per kilogram food consumed than is normal for sheep. Similar results were obtained in later experiments (Maloiy *et al.*, 1970). Like European cattle, they evidently need an abundant water supply, a point to be remembered when fencing in deer on a sometimes dry moorland paddock.

Preliminary measurements made by Brockway and Maloiy (1968) showed that the heat losses of fasting deer were considerably greater than those of sheep. When later measurements are included, the mean value for five hinds was 79 kcal/kg body weight $^{0 \cdot 75}$, compared with a value of about 60 kcal/kg$^{0 \cdot 75}$ for sheep, a difference of 30%. The rather high metabolic rate of deer was confirmed in much more detailed studies of three growing stag calves which otherwise resembled lambs in their ability to digest and metabolize a concentrate diet (Pollock, Webster & Smith, unpublished). Studies of the food requirements of dry and lactating hinds and of stag calves, combined with scrutiny of records of food consumption by penned mature deer extending over some years, indicate that the maintenance requirement for dry food or for digestible organic matter is some 30% higher in deer than in sheep of the same body weight (Arman, Kay, Goodall & Pollock, unpublished). Although this suggests that, weight for weight, deer may need more to eat at pasture than sheep, the maintenance requirement of housed animals is only one of the factors influencing the energy requirements of the grazing animal. Other important considerations are muscular activity, coat insulation, and the effectiveness of sheltering behaviour. Until each of these factors is assessed, or a more practical trial is undertaken, it is not possible to assess the comparative dietary needs of grazing deer and sheep.

Lactation and calf growth

If the growth potential of an animal from birth is to be examined, a logical starting point is a study of lactation. Dr. Pamela MacDonald, seconded to the Rowett Institute from the Nature Conservancy, recorded the volume and composition of the milk produced by tame hinds that were housed indoors (Arman, Kay, Goodall & Sharman, 1974). Milk yield during 24-h periods was measured by weighing the calf before and after each of the brief periods it was allowed to suck its dam. This gave rates of production in five well-fed hinds that reached peaks of from 1400–2000 g/day, one to five weeks after parturition, and declined fairly suddenly about two months later. The calves grew rapidly, gaining 250–350 g/day when wholly dependent on their dams. A poorly fed hind produced less than half this amount of milk, and her calf gained little weight until creep feed was made available to it. As shown in Table 11:1, the milk of these deer was very rich, containing in early lactation 8·5% of fat and 7·1% of crude protein, in late lactation 13·1% of fat and 8·6% of crude protein; lactose remained constant at about 4·5% throughout lactation. This milk is at least as rich as ewe's milk, and its composition explains the greater success achieved when captured calves were raised on ewe-milk substitute rather than concentrated cow-milk substitute (Chapter 5). The substantial increase in fat concentration, and the lesser increase in protein, that occurs as lactation proceeds appears to be a characteristic of the deer family; only relatively slight changes are observed in cattle or sheep.

Milk samples taken from Rowett and Glensaugh hinds have been examined by starch-gel electrophoresis to see if two variants of the whey protein, β-lactoglobulin, were present, as is commonly the case in cattle (E. I. McDougall, unpublished). Two of the samples did show a double zone for the protein, indicating two variants, but 16 others showed only a single zone.

The study of the Rowett hinds also provided the opportunity to make detailed observations of parturition and of the behaviour of dam and calf towards each other, albeit in quite abnormal surroundings (Arman, 1974). Points of interest that emerged were the grooming behaviour shown in varying degree by all the hinds towards their calves as a result of which part or most of the faeces and urine voided by the calf during the first few weeks of life was consumed by the dam, the often determined rejection by the underfed hind of her hungry offspring, and the willing fostering of a hastily substituted calf by a dam that gave birth to a dead calf.

Table 11:1 **Chemical composition of hind milk at various stages of lactation (from Arman, Kay, Goodall and Sharman, 1974)**

g/100 g milk	Stage of lactation days 3–30	31–100	101+	Average values for cow milk
Total solids	21·1	23·5	27·1	12·7
Fat	8·5	10·3	13·1	3·7
N × 6·38	7·14	7·63	8·59	3·5
Casein N × 6·38	5·71	6·16	7·03	3·3
Whey protein N × 6·38	1·13	1·23	1·34	
Non-protein N × 6·38	0·30	0·24	0·22	0·21
Lactose	4·45	4·45	4·46	4·8
Ash	1·18	1·11	1·10	0·7
Ca	0·22	0·22	0·25	0·12
P	0·22	0·18	0·19	0·10
Mg	0·018	0·018	0·022	0·012
Na	0·033	0·037	0·035	0·058
K	0·12	0·13	0·12	0·14
Cl	0·070	0·071	0·080	0·10

Two groups of five calves were weaned at 7·5 weeks of age from bottle-given milk on to a barley-based concentrate diet plus 10% of roughage (Arman, Kay & Goodall, unpublished). One group received this *ad libitum*, and the other was rationed to about 70% of the *ad libitum* intake. In the first group the stag calves gained about 330 g daily and the hind calves about 250 g daily, giving conversion ratios of dry food eaten to gain in body weight of from 2·7 to 3·8. The restricted group grew more slowly, though they compared favourably with most wild calves. A stag and hind calf from each group were slaughtered when they reached 48 kg and their carcases were dissected and analyzed. What was immediately striking was how remarkably lean these animals were, despite their rich diet (Table 11:2); their carcases contained only 4·8–5·6% of fat (16–20% of carcase dry matter), no more than half as much as would be found in a lamb raised on this diet. The characteristic leanness of wild deer is evidently not simply a result of poor nutrition.

In another trial, the growth rate of five stag calves given the same concentrate diet was recorded for a much longer period—about 2 years—and the effect of castration at 6 weeks of age was also studied in three of the calves (Arman, Kay & Goodall, unpublished). Growth rates and food conversion efficiencies were rather poorer than observed in the previous trial, but nevertheless the size of the animals when 26 months old was impressive, the two intact stags weighing 161 and 185 kg and the two surviving castrates (the third having been slaughtered for carcase analysis at 48 kg) weighing 107 and 112 kg. This is much more than the stags at Glensaugh weighed at this age, 80–100 kg. As the Rowett and Glensaugh animals were all taken without selection from the same mixed stock, it appears that the Glensaugh animals have not realised anything like their full genetic potential for growth at an age when they seem to have nearly reached their mature size and weight.

Figure 11:2 shows the food intake and weight changes of two of the Rowett stags. Once the intact stag began to approach sexual maturity, in his first summer, he did much better than the castrate. The latter retained the appearance of a calf, growing neither antlers nor a mane, and developed none of the aggressive rutting behaviour of the intact animal. He was mistaken for a hind at first glance even by experienced people. The marked seasonal fluctuation in appetite, and consequently in growth rate, that has been referred to earlier, is clearly seen. This occurs, it must be emphasized, in spite of the fact that a uniform diet was freely available to the deer throughout the experiment. This seasonal variation is quite evident in the castrate, though not as much as in the intact stag, and so cannot be wholly due to the enormous fluctuation in the secretion of testicular hormones that occurs during the year (p. 47). A similar but rather less marked variation in appetite is seen in hinds. These intriguing observations gave rise to a further experiment which is described later (p. 85).

Table 11:2 **Characteristics of the carcases of domestic and wild deer (Fowler, MacDonald & Pennie, unpublished), compared with those of cattle and sheep (Callow, 1947, 1948, 1958; Callow & Searle, 1956)**

Animal	Age (years, months)	Male Female Castrate	Diet	Liveweight (LW) (kg)	Empty body weight (kg)	Dressed carcase weight (DCW) (kg)	Dressed carcase (% of LW)	Fat, abdominal+ subcutaneous (% of DCW)	Boned-out first class meat (% of DCW)	Second class meat (% of DCW)	Remainjng carcase bones (% of DCW)
Deer											
Calves:											
210	0, 6	F	Concentrates, *ad lib.*	48	43	29	60	7	56	22	15
213	0, 5	M	Concentrates, *ad lib.*	49	44	28	58	5	56	23	16
211	0, 11	F	Concentrates, restricted	47	39	24	52	2	63	19	16
221	0, 7	M	Concentrates, restricted	48	43	27	55	4	59	22	15
Wild hinds:											
GF	ca. 3, 7	F	Good condition	60 (est.)	55	31	52	6	57	23	14
IE	ca. 10, 8	F	Poor condition	42 (est.)	38	27	65	1	56	25	18
Domestic hinds:											
Asali	2, 3	F	Grazing, Glensaugh	73	61	40	55	4	61	21	14
Rhum 5	5, 3	F		100	79	51	51	10	56	20	14
Domestic stags:											
Jitu	3, 3	M	Dried grass, *ad lib.*+concentrates	192	178	110	57	15	45	28	12
S11	2, 3	M	Concentrates, *ad lib.*	160	148	94	59	15	49	23	12
S7	2, 5	M	Concentrates, *ad lib.*	157	153	102	65	6	58	24	12
Cattle											
Hereford × Shorthorn	1, 7	C	Silage, *ad lib.*+concentrates	500	410	275	55	12	46	28	14
Sheep											
Suffolk cross	0, 4	M	Intensively fed	46	42	22	47	22	39	23	16
Cheviot	0, 5	F		37	33	17	46	22	40	22	16

Notes:
Liveweight—deer not fasted before slaughter.
Empty body weight—liveweight less gut contents.
Dressed carcase—liveweight less head, hocks, skin, blood, digestive tract, liver, lungs and heart.
First class meat—boned-out haunch, shoulder and back.
Second class meat—neck, thorax and most of backbone.
Carcase bones—remaining bones and tail.

Figure 11:2 **Seasonal changes in the rate of growth and food intake of an entire stag (S7) and a castrate stag (S12) given a concentrate diet *ad libitum***

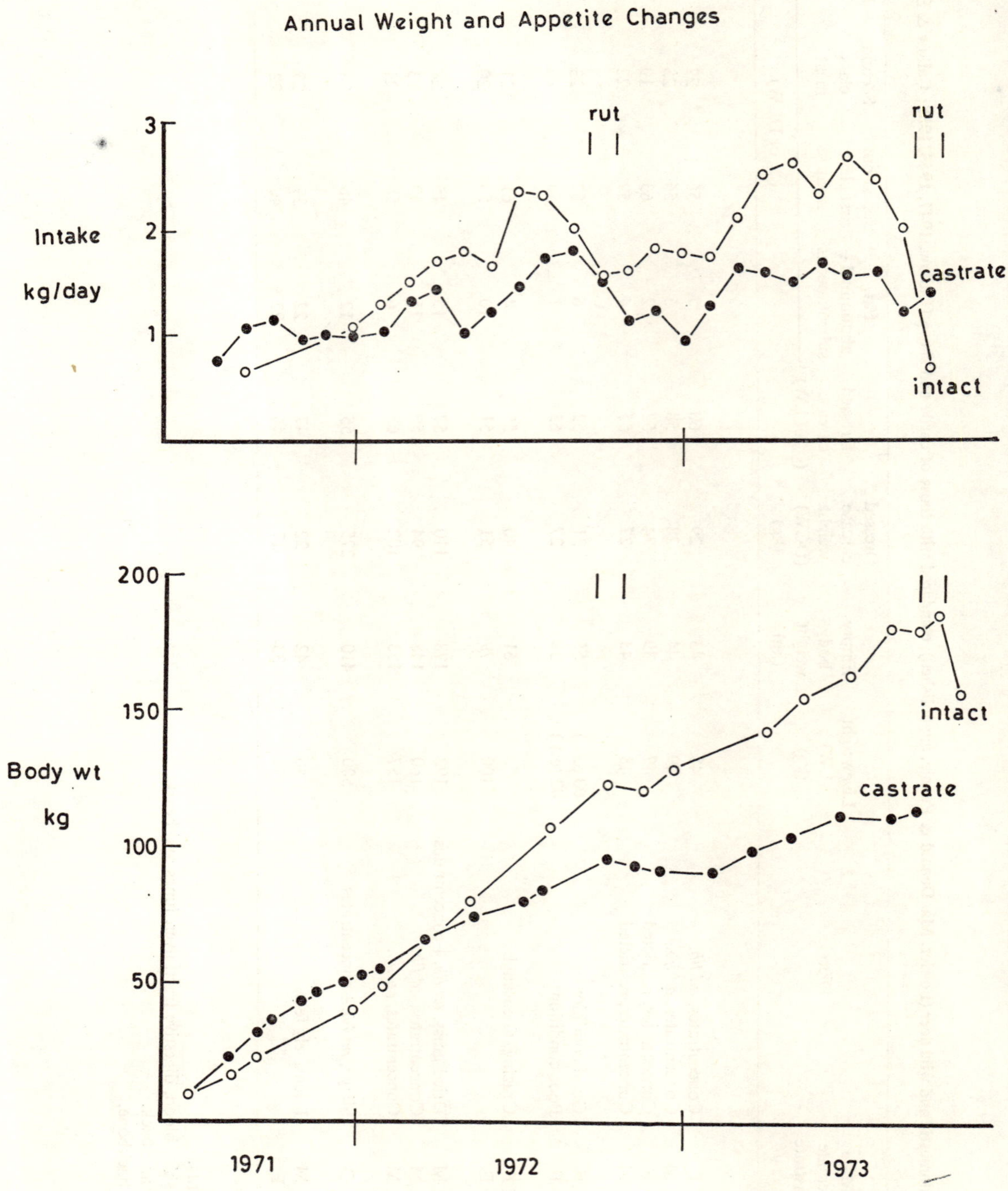

The composition and flavour of deer carcases

During the course of work at the Rowett Institute, deer that have been killed at the end of experiments have been subjected to the same type of carcase assessment as is commonly used for cattle and sheep. In addition, a proportion have been analysed chemically to measure the amounts of water, protein, lipid, energy and ash constituents in the body. As the farming enterprise develops some of the animals produced at the farm will also be subject to these procedures. Some typical data obtained by V. R. Fowler, P. MacDonald and K. Pennie (unpublished) are summarized in Table 11:2 where comparisons are made with cattle and sheep reared under comparable conditions. Data for wild hinds are also included.

The Table shows that deer calves 210 and 213, which were given high-quality diets and killed at about 48 kg liveweight when about 6 months of age, had a killing-out percentage (weight of dressed carcase as % of liveweight) of about 59%, and that some 56% of the dressed carcase could be regarded as first-class meat (boned-out haunch, fore-limb and back). Deer 211 and 221 were fed so that they grew about half as fast as deer 210 and 213, and when killed at the same liveweight they were slightly leaner and so yielded a slightly greater percentage of first-class meat. Intensively fed sheep, which have similar empty body weights to these deer at conventional slaughter weights, have lower killing-out percentages, about 47%, and owing mainly to their fat content, only some 40% of the dressed carcase consists of first class meat. The first-class meat obtained from a lean young deer carcase is 33% of its empty-body weight. In a sheep of similar size the amount of first-class meat obtained is only 18% of empty-body weight.

The nature of this first-class meat has to be considered. The carcases of the deer calves are extremely lean and contain very little intra-muscular fat. This is evident in the data given in Table 11:3 where the chemical compositions of the carcases are shown. Sheep and cattle carcases, on the other hand, contain quite large amounts of fat within and between the muscles. Plate 14 shows how very obvious this difference in fat content is in chops taken from young deer and sheep carcases at comparable weights.

The two mature domesticated hinds, Rhum 5 in good condition and Asali in hard condition, had a larger proportion of gut contents than the calves, leading to a lower killing-out percentage. However, they were little fatter and yielded little or no poorer proportion of first-class meat than the calves. The wild hinds represent two extremes, one in good condition and the other very poor and with next to no fat on her. The yield of first-class meat from these animals was again greatly superior to that found in sheep.

The three mature domesticated stags, Jitu, S7 and S11, are of considerable interest. Jitu was given dried grass with a concentrate supplement and S7 and S11 a barley-based concentrate *ad libitum* and it will be noted that they had laid down a lot of fat. It seems that well-fed stags, at least, may grow as fat as cattle or sheep slaughtered at normal market weights but they deposit this fat only when skeletal growth is complete. The dressed carcase still formed a greater proportion of body weight than is found in cattle and considerably greater than in sheep, but the proportion of first-class meat in this carcase was little greater than in cattle.

Stag 11 was killed on 17th September at the beginning of the rut and before losing much condition. Stag 7 was killed on 5th November at the end of the rut. His appetite had been very low for several weeks and he had lost 28 kg body weight since topping 185 kg on 18th September. By comparison with S11 it seemed that about two-thirds of this fall in weight was due to loss of fat, taken equally from subcutaneous, abdominal and intra-muscular sites (Table 11:3), and one-third was due to reduction in the weight of the digestive tract and contents. These two factors together were responsible for his high killing-out percentage, 65%, which is typical of rutting stags.

Another noticeable feature of the deer carcases was that they were darker in colour than those of sheep or cattle. This observation in hand-fed well-butchered deer suggests that the characteristically dark colour of wild venison is not due wholly to high blood content and hard condition.

Besides assessing the 'domestic' deer carcase from the butcher's point of view, some preliminary steps have been taken to consider its attractiveness to the housewife. Early tests of the cooking properties of calves butchered at the Rowett Institute (including calves 210 and 213, Table 11:2) were conducted by Miss J. Butchart of the College of Domestic Science, Robert Gordon's Institute of Technology, Aberdeen. No difficulty was encountered with roasting this meat although wild venison has the reputation of being an awkwardly 'dry' meat. The meat was very tender and quite acceptable, especially to people who did not like the flavour of wild venison; those who did like wild venison thought the flavour of the calves, especially the hinds, was rather disappointing.

More extensive trials were later kindly made by Mr. D. M. Oxtoby and Mr. H. Steven of W. A. Baxter & Sons, Ltd., Fochabers. Experienced tasting panels were offered various sorts of canned and roast venison, prepared from sides of three mature Rowett stags (including Jitu and S11, Table 11:2) that had been normally stunned, bled and deep frozen, and from a wild stag shot on the hillside and not frozen. Members of the panel that were accustomed to eating wild venison found the Rowett stags moderately to very acceptable but lacking in gamey flavour, and so they rated them less attractive than the wild stag. Panel members not accustomed to venison mainly agreed the wild stag had the most gamey flavour but this made most of them prefer the Rowett meat. As Mr. Oxtoby remarked, these tests indicate that even in the North of Scotland the gamey flavour of venison is strange to many people and by and large is not liked, a factor in favour of farmed venison being sold in the U.K. but probably against it in continental markets.

Table 11:3 Concentrations of total extractable fat in the carcases of deer (Fowler, MacDonald & Pennie, unpublished) compared with those in cattle and sheep (Callow, 1947, 1948, 1958; Callow & Searle, 1956)

Animal	Age (years, months)	Male Female Castrate	Diet	Dressed carcase weight (kg)	Extractable fat (g/100 g wet weight): In dressed carcase less subcutaneous and abdominal fat	In muscle
Deer						
Calves:						
210	0, 6	F	Concentrates, *ad lib.*	29	7·3	2·0
213	0, 5	M	Concentrates, *ad lib.*	28	5·8	1·8
211	0, 11	F	Concentrates, restricted	24	1·3	0·6
221	0, 7	M	Concentrates, restricted	27	6·7	1·9
Wild hinds:						
GF	ca, 3, 7	F	Good condition	31	8·7	2·4
IE	ca. 10, 8	F	Poor condition	27	2·7	1·3
Domestic stags:						
Jitu	3, 3	M	Dried grass, *ad lib.*+concentrates	110	15·0	6·8
S11	2, 3	M	Concentrates, *ad lib.*	94	23·2	10·5
S7	2, 5	M	Concentrates, *ad lib.*	102	9·5	3·7
Cattle						
Shorthorn 10	1, 9	C		303	24·5	5·8
Shorthorn 12	2, 0	C		336	19·8	5·2
Sheep						
Suffolk 178	0, 5	F		25	24·0	6·6
Suffolk 174	0, 6	C		23	20·6	7·0

Chemical composition of deer fat

The flavour and texture of meat are affected not only by the quantity of fat it contains and its distribution, but also by the chemical composition of the fat. Chemical analyses of the fat of various mammals, including red deer and other species of deer, made by G. A. Garton and his colleagues at the Rowett Institute are, therefore, relevant to the deer farming venture.

The fatty-acid composition of triglycerides extracted from perinephric, omental and subcutaneous adipose tissue of six wild red deer was determined by Garton & Duncan (1971). The animals, which were shot in Glen Esk, Angus, comprised two non-pregnant hinds, two pregnant hinds and two lactating hinds. As in adult sheep and cattle, the perinephric triglycerides of the non-pregnant, non-lactating hinds are characterized by the presence of saturated fatty acids and, in particular, of palmitic and stearic acids which together account for some two-thirds of the total fatty acids. In the sheep, the triglycerides of omental and perinephric tissue closely resemble each other in fatty-acid composition, but this is not so in the deer whose omental triglycerides, while still retaining the same degree of unsaturation as those of perinephric tissue, are notable for their comparatively high content of myristic and palmitic acids accompanied by a correspondingly lower content of stearic acid. As is usual in both ruminant and non-ruminant animals, the subcutaneous triglycerides of the red deer are more unsaturated and so have a lower melting point, appropriate in a tissue exposed to the cold, than the deeply-placed abdominal fat depots.

Though pregnancy does not appear to be associated with changes in the fatty-acid composition of the triglycerides, the demands of lactation are reflected in marked changes, principally an increase in the proportions of myristic, palmitic and palmitoleic acids at the expense of stearic acid.

Photoperiodic control of moulting, appetite and sexual activity

Seasonal changes in the coat of the deer were first studied in four of the hinds housed at the Institute. Samples of the coat and small snips of skin were removed from the flank at monthly intervals for 2 years and changes in the nature of the coat and the activity of hair follicles were recorded (Ryder & Kay, 1973). Wild red deer have a coat comprising stiff coarse guard hairs about 5 or 6 cm long which in winter overlie a fine woolly undercoat, and it changes seasonally from a red summer colour to a greyer winter coat. In the Rowett hinds the hair and wool follicles were dormant in the depth of winter but follicle activity and hair growth, mainly of the guard hairs, appeared during February and continued until about May. Between March and June the winter coat was moulted, the deer becoming very scruffy in appearance before the last patches of old coat were cast to reveal a glossy red summer coat containing almost no wool. From May to August there were irregular minor cycles of follicle activity and dormancy, culminating in a period of intense activity of both hair and wool follicles from August to October as the winter coat was grown. Moulting of the summer coat occurred during September and October but was not at all obvious to the casual observer as it was a gradual process and, since little wool was cast, the hairs fell out individually rather than in matted clusters. Moulting of an old coat did not occur until the new coat was well grown, so that the deer were never left without good insulation.

A similar study on a group of stags and hinds at Glensaugh was completed in December 1973 (Ryder & Kay, unpublished). Coat changes in the grazing animals followed the general pattern observed in the Institute animals, although follicle activity and moulting tended to occur a month or two later and to show less variation between animals or years. The variability in the Rowett deer may have been due to the effect of using some artificial light in the deer pens in winter, as it became clear later that day length has an effect on the moulting cycle, although nutrition and environmental temperature may also have influenced coat growth to some extent.

The observations on seasonal changes in appetite, antler growth, rutting behaviour and coat growth, that have been recorded above and in earlier chapters, suggested that day length may play an over-riding role in the timing of these changes. If so, it means that these physiological changes are not directly stimulated by the environmental hazards to which they are adaptations, namely the quality and availability of forage and the severity of the climate, but by the underlying and entirely regular changes in day length.

In order to find what physiological changes were associated with day length, four stag calves were placed in blacked-out pens in September 1972 and were subjected to an artificial lighting regime in which changes in 'day' length followed a 6-month cycle, rather than the natural 12-month cycle, with 'midsummers' occurring in December and June and 'midwinters' in September and March (Pollock, Kay & Goodall, unpublished). Normal annual temperature changes were reduced but not eliminated. The stags were offered *ad libitum* a cereal-based concentrate diet plus 10% dried grass. Figure 11:3 illustrates some of the results. The stags began to grow antlers precociously in December and subsequently have shown cycles of antler casting, growth and fraying, of testis size and plasma testosterone concentration and of rutting behaviour which have followed a clear-cut 6-month period. Appetite also varied in a marked manner, falling abruptly at the time of the rut and not recovering again until the next phase of antler growth, with consequent effects on body weight. Changes in the coat growth cycle were more complex and did not show complete adaptation to the imposed cycle.

These observations are interesting from many points of view. They show how neatly deer have become adapted to the seasonal vicissitudes of their natural environment; possibly deer, despite their lean nature, are better able to withstand short rations in winter than our present domestic ruminants and if so, this might allow considerable savings in expensive winter keep. They suggest ways in which it may be possible to encourage appetite and growth in housed deer. They provide a challenge and perhaps a suitable choice of experimental animal for unravelling the neuroendocrine mechanisms which underlie the responses to variations in photoperiod. What is more we may have sadly neglected a line of research which could be of considerable practical importance in the management of conventional farm animals.

Haematology

Ideally, for obtaining reproducible results, the patient should be calm at the time of blood sampling for haematology but, of course, calmness is a state difficult to produce in deer that are being handled, even if they are tame. As in any deer farming enterprise the animals are likely to be somewhat excited when restrained for blood sampling, it was decided to take the opportunity at the deer farm when the deer were being gathered for weighing and measuring each month, to take blood samples from certain individuals and carry out determinations for haemoglobin (Hb), packed cell volume (PCV) and red and white cell counts (RCC and WCC). Deer kept in pens at the Rowett Institute were also sampled.

The deer were either caught by hand as they emerged from the weighing crate or in their pens, or they were manhandled into a metal restraining crate. Blood was withdrawn from a jugular vein through a 20-g needle into an evacuated tube containing a trace of dried heparin. The samples were refrigerated overnight (seven deer from the deer farm) or were examined within a few hours of sampling (four deer at the Rowett Institute). Some 160 observations were made between July 1971 and August 1973 from deer one month to two years old. Hb was determined by a cyanhaemoglobin colorimetric method, PCV by microhaematocrit centrifugation at 12,000 g for 5 minutes, and RCC and WCC by using a Coulter Counter.

There were no differences between the results obtained from deer at the deer farm and at the Rowett Institute, nor was there any evidence of systematic changes with time. There were no sex differences (male, female or male castrate castrated at six weeks of age). Table 11:5 gives the overall means and the range for each of the parameters. In general the values found for Hb and PCV were considerably higher than those for normal sheep, goats and cattle. RCC's were about the same as those accepted for sheep and goats and very young cattle, while WCC's were lower than in sheep, goats and cattle. In cattle the red cell parameters Hb, PCV and RCC decline with age. The youthfulness of our deer may in part account for the relatively high values but it is more likely that they and the low WCC are inherent characteristics of deer.

Table 11:5 **Haemogram of 11 deer 1 month to 2 years old, ca. 160 observations**

		Mean	*Range*
Hb	g/100 ml	16·2	12·4—20·6
PCV	%	44·9	33–55
RCC	10^6/cmm	11·3	7·1—16·5
WCC	10^3/cmm	4·4	1·7—8·6

Table 11:6 **Number of phenotypes and gene frequency Tf_{ab}, for serum transferrins found in red deer at Glensaugh**

Phenotype	ab	abc	bc	Tf_{ab}
Observed number	65	26	9	0·78
Expected number	60·8	34·3	4·8	

Figure 11:3 **'Seasonal' changes in food intake and body weight, in antler growth and in testis size in a stag subjected to a 6-monthly cycle of day-length changes**

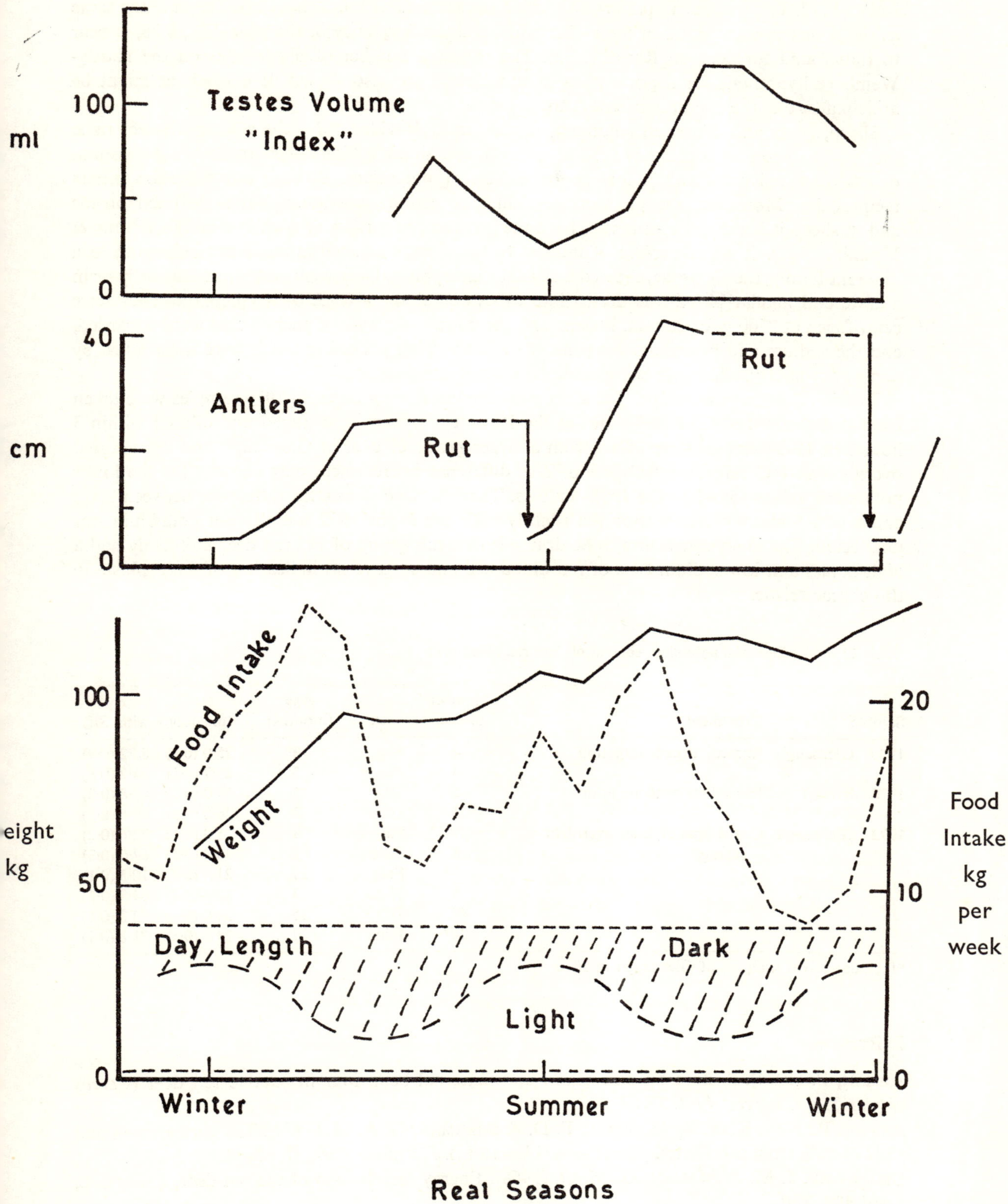

Studies were also made of two characteristics of the blood unlikely to be affected by the alarm caused by restraint and blood sampling, the serum transferrin patterns and the sodium and potassium content of the red blood cells.

The transferrins are a group of closely related proteins which can be separated by starch-gel electrophoresis and labelling with radioactive ^{59}Fe. The patterns of transferrins show characteristic and inheritable differences between individuals and races of a species. McDougall & Lowe

(1968) examined blood obtained from deer shot on Rhum and near Braemar and found that the transferrin patterns appeared to comprise a system of three common phenotypes attributed to two alleles. Subsequently samples of serum were obtained from 100 deer at Glensaugh in September 1972. The transferrin patterns observed are summarized in Table 11:6. They are the same as the common ones found in the earlier work and the frequency of the gene Tf_{ab}, 0·78, is near to that found for deer on Rhum (0·73). The expected number of phenotypes on the Hardy-Weinberg hypothesis for a population in equilibrium are also shown. It is clear, as might be anticipated, that the Glensaugh herd is not in genetic equilibrium.

Sheep, goats and cattle, all belonging to the family Bovidae, have the curious characteristic that while some individuals have red blood cells containing a high concentration of potassium, others have red cells with a low concentration of potassium. Sodium concentration differs reciprocally. Moreover, the proportion of high-potassium individuals varies between breeds and in sheep it is the hill breeds that have the greatest proportion of such individuals (Evans & Mounib, 1957). It was therefore of interest to see if red deer, hill animals but belonging to a different family, the Cervidae, showed a similar dimorphism. Deer were restrained manually or in a crate and jugular blood was drawn into tubes containing a trace of dry heparin. They were centrifuged at 2700 g for 20 min. Plasma and the uppermost layer of packed cells were discarded, and the sodium and potassium contents of the underlying packed red cells were determined by flame photometry, allowance being made for mutual interference.

Table 11:7 shows the results. These are presented in three groups. In 1970 samples were taken from deer housed at Glensaugh and at the Rowett and were centrifuged and diluted within 3 hours. In 1972 samples were taken from a larger group of deer at Glensaugh and were stored overnight at 1°C before centrifugation. This difference in treatment may account for the lower potassium values found in the 1972 samples. There was no consistent difference between stags and hinds. Sodium concentration fell slightly with age in the 1972 samples but potassium was unaffected. The main conclusion to be drawn is that this group of 113 red deer uniformly had a high-potassium, low-sodium type of red blood cell and so gave no indication of dimorphism in this characteristic.

Table 11:7 **Sodium and potassium content of deer red blood cells**

Sample		Treatment	Number of deer	Sex	Age (months)	Na^+ (m.equiv/l	K^+ with SE)
1970	Glensaugh	Immediate centrifugation	4	Stag	3	25·0 (0·9)	87·2 (0·9)
			4	Hind	3	25·7 (0·1)	84·2 (0·8)
1970	Rowett	Immediate centrifugation	5	Stag	3	29·0 (0·5)	85·8 (0·2)
			5	Hind	3	26·8 (0·2)	85·8 (0·1)
1972	Glensaugh	Centrifugation after overnight storage	7	Stag	3	25·9 (0·1)	72·6 (0·2)
			8	Stag	15	25·2 (0·1)	67·0 (0·3)
			8*	Stag	27	21·0 (0·6)	70·5 (0·1)
			8	Hind	3	33·1 (0·2)	64·8 (0·1)
			48	Hind	15	30·0 (0·7)	73·2 (0·1)
			24*	Hind	27	23·9 (0·5)	71·8 (0·1)

*Includes calves sampled at Glensaugh in 1970

References

Arman, P. (1974). A note on parturition and maternal behaviour in captive red deer (*Cervus elaphus* L.). *J. Reprod. Fert.* **37,** 87–90.

Arman, P., Kay, R. N. B., Goodall, E. D. & Sharman, G. A. M. (1974). The composition and yield of milk from captive red deer (*Cervus elaphus* L.). *J. Reprod. Fert.* **37,** 67–84.

Brockway, J. M. & Maloiy, G. M. O. (1968). Energy metabolism of the red deer. *J. Physiol., Lond.* **194,** 22*P*.

Callow, E. H. (1947). Comparative studies of meat. I. The chemical composition of fatty and muscular tissue in relation to growth and fattening. *J. agric. Sci., Camb.* **37,** 113–131.

Callow, E. H. (1948). Comparative studies of meat. II. The changes in the carcass during growth and fattening, and their relation to the chemical composition of the fatty and muscular tissues. *J. agric. Sci., Camb.* **38,** 174–199.

Callow, E. H. (1958). Comparative studies of meat. VI. Factors affecting the iodine number of the fat from the fatty and muscular tissue of lambs. *J. agric. Sci., Camb.* **51,** 361–369.

CALLOW, E. H. & SEARLE, S. R. (1956). Comparative studies of meat. V. Factors affecting the iodine number of fat from the fatty and muscular tissue of cattle. *J. agric. Sci., Camb.* **48,** 61–73.

EVANS, J. V. & MOUNIB, M. S. (1957). The survey of the potassium concentration in the red blood cells of British breeds of sheep. *J. agric. Sci., Camb.* **48,** 433–437.

GARTON, G. A. & DUNCAN, W. R. H. (1971). Fatty acid composition and intramolecular structure of triglycerides from adipose tissue of the red deer and the reindeer. *J. Sci. Fd Agric.* **22,** 29–33.

MCDOUGALL, E. I. & LOWE, V. P. W. (1968). Transferrin polymorphisms and serum proteins of some British deer. *J. Zool., Lond.* **155,** 131–140.

MALOIY, G. M. O. (1968). The physiology of digestion and metabolism in the red deer (*Cervus elaphus* L.). *Ph.D. thesis: University of Aberdeen.*

MALOIY, G. M. O. & KAY, R. N. B. (1971). A comparison of digestion in red deer and sheep under controlled conditions. *Q. Jl exp. Physiol.* **56,** 257–266.

MALOIY, G. M. O., KAY, R. N. B., GOODALL, E. D. & TOPPS, J. H. (1970). Digestion and metabolism in sheep and red deer given large or small amounts of water and protein. *Br. J. Nutr.* **24,** 843–855.

RAZZAQUE, M. A., TOPPS, J. H., GOODALL, E. D. & KAY, R. N. B. (1973). Metabolism of nucleic acids by sheep and red deer. *Proc. Nutr. Soc.* **32,** 59A.

RYDER, M. L. & KAY, R. N. B. (1973). Structure of and seasonal change in the coat of red deer (*Cervus elaphus*). *J. Zool., Lond.* **170,** 69–77.

Chapter 12. **Prospects and future plans**

Some of the investigations described in previous chapters have progressed so far that they permit tentative conclusions; others are obviously not so well advanced. On the basis of this work it is not possible to state categorically that deer farming as initially conceived is as yet capable of being developed commercially with certainty of success. Nevertheless, what has been accomplished enables a closer and more informed examination of the ways in which commercial development could take place than was possible when experimental farming of the red deer began in 1970. Three broad groups of problem can be considered; firstly, those concerned with methods of establishing deer farms, secondly, those related to the control of the stock and thirdly, those which are pertinent to the likely productivity of deer farming enterprises. In each of these broad divisions the two Institutes are planning to undertake further work; any who are contemplating commercial development are advised to await the result rather than to assume that particular alternatives which have not been tested are practicable.

Herd establishment

The herd at the farm was established by catching new-born calves in the wild, rearing them artificially, reinforcing their tameness by hand-feeding them during their first winter and by handling them at monthly intervals. There is no doubt that this method of establishing a herd resulted in a very tame stock of hinds and when the amenability of this stock is compared with that of captured adults kept closely confined in pens for many months and subsequently released on the farm, it is clear that an initial establishment based on a hand-reared stock is the more satisfactory. Furthermore, provided that the calves born to this initial stock are handled with their dams, weaned before the rut, are hand-fed during their first winter and subsequently gathered at monthly intervals, the second-generation stock can be managed, although not as easily as can conventional farm livestock.

Whether commercial farms can be established in this way on any scale is open to doubt. The Red Deer Commission, which has statutory responsibility to deal with all matters concerning red deer, gave the two Institutes their special permission to capture calves for experimental purposes during the close season and such permission is not likely to be given readily to intending deer farmers. This suggests that any major commercial development of deer farming based on initial stocks of captured calves is unlikely. Rather development may well occur through the domestication of naturally suckled calves derived either from a confined wild stock, park stocks or from small nucleus herds of domesticated deer such as that at the Glensaugh farm. If this is the pattern which will ensue, commercial development will be slow and it is to be wondered whether the limited experience obtained at the farm about the second generation of calves is a sufficient basis for planning. For this reason, and commencing in summer 1974, the two Institutes will establish a second farm where direct comparisons will be made between a naturally reared stock derived from the farm hinds and an artificially reared stock derived from captured calves. It is intended at this farm to reduce the frequency of handling to one more appropriate to a commercial system.

It should be pointed out that deer farming of an entirely different nature to that envisaged here has commenced in New Zealand and this involved establishment with adults. Farming there simply consists of containing a wild stock in a large fenced area. Wild adult deer in remoter areas have been stalked and immobilized using immobilizing drugs to be transported to these 'farms' by helicopter and lorry, and in New Zealand, planning the essential features of centralized slaughter and eventual domestication do not appear to have been given the prominence accorded them in the Scottish study, although careful attention has been given to the hygienic processing of carcases. Obviously, the fencing of parts of existing deer forests in Scotland could result in similar types of enterprise to those being started in New Zealand. Such undertakings, however, still entail treating the deer as a wild animal, and many of the husbandry practices adopted at the experimental farm are hardly feasible in such a context. It is, however, possible that such a containment of wild stocks could facilitate the eventual establishment of farms in the true sense of that term since they could supply weaned calves that could be tamed.

Management and control of the stock

Control of a deer stock includes all aspects of their management, including disease control and grazing management. At the farm simple and cheap fencing has been devised to contain the stock, and very effective handling facilities have been built which have worked well. Experience has shown that there are few problems in moving hand-reared stock from one area of the farm to another. Success in these aspects of control is clearly related to the tameness of the animals, and it is not yet known whether with smaller expenditures of time and care a naturally weaned stock can be managed in a similar way. The calves born in 1973 to Glensaugh hinds were handled to some extent while with their dams and after weaning were hand-fed through winter, yet they are considerably less easy to manage than their bottle-reared dams. The establishment of the second farm should answer this question. The fences and handling facilities described in Chapter 4 should be suitable for any commercial enterprise since fencing needs are not greatly affected by these uncertainties. Methods of stock movement might well require modification.

Control of breeding at the farm has not progressed very far. Experience has shown that the movement of mature stags at the time of the rut is a hazard to those concerned and it is not yet known how far the number of intact stags, and hence the magnitude of the hazard, can be reduced without affecting the reproductive performance of the herd as a whole. The preliminary finding discussed in Chapter 7 that an isolated stag is not effective as a sire even when given a small group of hinds, if confirmed, necessarily entails keeping more stags than was envisaged originally, but the size of mating groups and the optimal ratio of hinds to stags in the herd has not yet been ascertained.

For any meat-production enterprise based on deer, unlike sporting enterprises, the number of mature stags should be kept to a minimum so to maximize the production of carcases each year. Some have suggested that meat production and the sport of stalking are compatible in a farming context. This is doubtful for if the stags are managed in such a way that they can be moved at the time of the rut and can be inspected and subjected to veterinary treatment, though they might well provide excellent trophies (provided their antlers were not sawn off), they would hardly provide much challenge to the sportsman. Conversely the harvesting of meat from a wild population on a sporting estate, though it provides excellent sport, cannot be regarded as meeting the requirements of a modern meat-producing industry.

Clearly further development of deer farming will entail further investigation of the breeding of the stock. In addition, it is hoped that an assessment can be made of the possibility of their genetic improvement, for it is not only likely that growth rate and ability to thrive will show genetic variation, but so also will those various attributes of behaviour which combine to produce a manageable animal.

The experience gained at the farm, which has been summarized in Chapter 10, shows that no major problems of disease and parasitism have emerged to suggest that keeping of deer at a high density would lead to serious declines in the health and well-being of the stock. Three and a half years of total operation, only the last of which can be regarded as beginning to approach a density of stocking which the herbage resources of the farm can support is, however, hardly an adequate basis on which to draw the more positive conclusion, which is required if deer farming is to become a commercial reality, namely that disease presents no problem even when stocking is at a rate that ensures maximal utilization of herbage available. In this respect, the success so far achieved may well relate in part to the methods used to establish the herd. The captured calves were removed shortly after birth from sources of infestation with parasites—though some admittedly did subsequently develop warbles. They were also removed from contact with adults of their kind. They did not graze at the farm until more than three months of age and the hill on which they were grazed had not been stocked with sheep during the previous two years. Both the land and the calves thus started with the advantage of low parasitic burdens, and, understandably, there was little evidence of parasitism in the herd at the end of the period reviewed.

Even though, in this short initial period, the stock has remained healthy, there is evidence that certain problems of parasitism and disease could emerge at some later time. Firstly, the irritation to the deer caused by the head fly (*Hydrotaea irritans*) is an obvious problem, as indeed it is in sheep in some areas, and the confinement of the deer means that a restraint is placed on their normal reaction to avoid the flies, namely to move to higher ground. Investigations of repellants are continuing. Secondly, it is possible that as the stocking rate increases lung and gastro-intestinal worm infestations will increase. Continued monitoring of the herd should show whether this will occur; strategic use of anthelminthics could well provide control. There is no virtue in the downward adjustment of stocking to prevent the emergence of parasitism if this entails an acceptance of poor use of herbage resources, particularly if medication can ensure a greater productivity. Thirdly, the major health concern may well prove to be the presence of reactors to the avian strain of mycobacterium that causes tuberculosis. Tuberculosis is certainly present in wild deer and it seems highly probable that wild birds play some role in its transmission. If hand-feeding—which attracts birds—results in contamination of the deer's feed with *M. avium* from infected droppings then it may well be that, quite apart from the greater density of stocking of deer on farms and the consequent greater risk of animal-to-animal transmission, the management practices on such farms will entail a greater infection rate with tuberculosis than occurs in the wild. The continuing tuberculin-testing of the stock at the main farm and at the second farm should provide a better appreciation of this risk.

Productivity

The general principles which determine the output of meat from a self-contained stock of deer were given in Chapter 2. There it was emphasized that the main determinant of total numbers of animals is the carrying capacity of the land during the winter, and that the annual meat output from them depends on the proportion of the total stock which are breeding hinds and their annual reproduction rate.

At the experimental farm total stocking rate has not yet approached the carrying capacity of the land, and on the evidence given in Chapter 9, it is thought that this capacity is about 150 adult deer or their equivalent as other age classes on 550 acres, that is one deer to 3·7 acres, provided that calves in their first winter are hand-fed supplementary feed, pregnant hinds are given additional feed before spring growth occurs, and hay is available for storm feeding. This could be an over-estimate of stocking rate, for winters in the three and a half years of the project have been milder than usual. These tentative estimates suggest that deer could be stocked on hill land at the same or a slightly lower rate than experience would suggest that hill sheep can be stocked. With these rates utilization of herbage by the deer would be greater than that by sheep.

No meat has yet been produced from the farm, but the studies of the carcases of wild and intensively reared deer given in Chapter 11 together with the records and observation of the growth of the foundation stock given in Chapter 6 suggest that animals kept continuously at the farm would reach a suitable weight for slaughter when 15–17 months of age. Yearling stags might well be slaughtered during August and September before the rut and yearling hinds in October-November. These times are immediately before weight loss or weight stability ensues and are also within the period when slaughter of red deer is permitted under the Deer (Scotland) Act. Stags would then weigh about 75 kg and hinds 65 kg. The experimental studies suggest alternatives. Since with high-quality feed calves can be reared to weights of 60 kg in about 7 months, it is possible to envisage removal of the calves from the farm when 2–3 months of age to be given better-quality feed either on lowland grass or in yards. Such a system would reduce winter stocking rates on the hill but would entail slaughtering calves in the close season. In neither system does it seem necessary to castrate the stags, for intact animals destined for slaughter would not be kept to ages greater than 15 months and there is so far no evidence that castration has any beneficial effect on early growth. Much additional work needs to be done and is planned to explore methods for producing carcases of good quality under different systems.

The farm also provides information of a limited nature about herd structure, fertility, mortality and culling when deer are treated as domesticated rather than wild animals. Hinds will calve at two years of age with a fertility of 90% and it would seem that older hinds are more fertile. Mortality up to weaning time is likely to be 10% or less. Length of productive life of hinds is not yet known but is expected to be an average of at least 5 years, implying that not more than 20% of the hind stock must be replaced each year. No experience of handling rutting stags older than 3½ years under farm conditions has been obtained and tentatively it is suggested that stags should be disposed of during the summer following their third rut, implying a replacement rate of breeding stags of 33% per annum. A proportion of the hinds and most of the stags which leave the herd would be killed for meat. Alternative strategies can be envisaged. Hinds, for example, might conceivably have a productive life averaging 8 years when the replacements required annually would be considerably reduced.

From such data calculations can be made of herd structure and meat output from a breeding herd of 100 hinds. Maximal herd size would occur in the late summer before the slaughter of meat stock. Then the herd would consist of 100 hinds, 85 calves, about 80 yearling stock some of which are destined to replace breeding hinds and stags, and 15 stags, 4 or 5 of which are four years of age and destined for slaughter. The meat crop would consist of 20 surplus yearling hinds, 35 surplus yearling stags, 10–15 culled hinds and 4 or 5 culled stags, equivalent to an output of over 5000 kg live weight or 3000 kg of dressed carcase from the herd. Mid-winter stocking would be 100 in-calf hinds, about 82 weaned calves and 15 stags more than one year of age. In terms of adult deer equivalents this winter stocking represents about 165 deer.

The above calculations are probably estimates of maximal potential and are limited by the accuracy of the initial estimates of fertility and mortality. Alternative estimates of these and alternative strategies of replacement of both hinds and stags have been examined by computer simulation to suggest that if the limit to the size of a herd is imposed by winter carrying capacity, meat stock should be slaughtered as early as possible and hinds and stags should be kept in the herd for as long as possible. Clearly the continuing observations on the farm will show whether predictions based on these initial estimates are in fact realised.

It is not realistic to cost the present experimental venture so to obtain an estimate of the economic viability of deer farming. Inputs of labour and technical and managerial skill have obviously been much greater than would be possible on any commercial farm. Even if the estimates of productivity are borne out by further trial at the farm, it is not known whether they can be applied to a more practical situation. The second farm, although planned to answer questions related to herd establishment, should provide information on the economics of farming since it is intended to minimize managerial care and experimental handling. Nevertheless, the calculations relating to output of meat together with the costs of fencing, the records of the amounts of supplementary feed supplied and estimates of land rentals and labour charges enable those who might contemplate farming the red deer to estimate its economic profitability in particular contexts.

Printed for Her Majesty's Stationery Office by David J. Clark Limited, Glasgow
Dd. 393252/3364 Gp. 221 K3 7/75